ROUTLEDGE LIBRARY EDITIONS:
URBANIZATION

Volume 9

URBANIZATION IN POST-APARTHEID SOUTH AFRICA

URBANIZATION IN POST-APARTHEID SOUTH AFRICA

RICHARD TOMLINSON

LONDON AND NEW YORK

First published in 1990 by Unwin Hyman Ltd

This edition first published in 2018
by Routledge
2 Park Square, Milton Park, Abingdon, Oxon OX14 4RN

and by Routledge
711 Third Avenue, New York, NY 10017

Routledge is an imprint of the Taylor & Francis Group, an informa business

British Library Cataloguing in Publication Data
A catalogue record for this book is available from the British Library

ISBN: 978-0-8153-8014-6 (Set)
ISBN: 978-1-351-21390-5 (Set) (ebk)
ISBN: 978-0-8153-7854-9 (Volume 9) (hbk)
ISBN: 978-1-351-23207-4 (Volume 9) (ebk)

Publisher's Note
The publisher has gone to great lengths to ensure the quality of this reprint but points out that some imperfections in the original copies may be apparent.

Disclaimer
The publisher has made every effort to trace copyright holders and would welcome correspondence from those they have been unable to trace.

URBANIZATION
IN POST-APARTHEID
SOUTH AFRICA

URBANIZATION IN POST-APARTHEID SOUTH AFRICA

Richard Tomlinson

London
UNWIN HYMAN
Boston   Sydney   Wellington

Published by the Academic Division of
Unwin Hyman Ltd
15/17 Broadwick Street, London W1V 1FP, UK

Unwin Hyman Inc.,
955 Massachusetts Avenue, Cambridge, MA 02139, USA

Allen & Unwin (Australia) Ltd,
8 Napier Street, North Sydney, NSW 2060, Australia

Allen & Unwin (New Zealand) Ltd in association with the
Port Nicholson Press Ltd,
Compusales Building, 75 Ghuznee Street, Wellington 1, New Zealand

First published in 1990

British Library Cataloguing in Publication Data

Tomlinson, Richard
   Urbanization in post-apartheid South Africa.
   1. South Africa. Urbanization
   I. Title
   307.760968

ISBN 0–04–445794–4

Library of Congress Cataloging in Publication Data

Tomlinson, Richard. 1952–
   Urbanization in post-apartheid South Africa / Richard Tomlinson.
      p.   cm.
   Includes bibliographical references.
   ISBN 0–04–445794–4
   1. Urbanization–South Africa–Forecasting.   2. Rural development–
   South Africa–Forecasting.   3. Population forecasting–South
   Africa.   4. Urban policy–South Africa.   5. Housing policy–South
   Africa.   I. Title.
   HT384.S6T66   1990   90–35926
   307.76'O968'01–dc20   CIP

Typeset in 10/11 point Bembo
Printed in Great Britain by Billing & Sons Ltd., Worcester

Contents

List of tables

List of maps

List of figures

Preface

A problem with writing about South Africa is that one has a tendency to litter the text with 'snigger brackets'. This is unavoidable since it is possible to lose one's reputation through carelessly referring to the present government's 'reforms' without so enclosing the term 'reform'. However, with a view to a visually simple and clear text, I have used quotation marks only on the first occasion that a sceptical interpretation seems appropriate. Yet snigger brackets alone will sometimes not suffice and further disclaimers may be necessary to establish that one despises racism and supports one-person, one-voue in a democratic South Africa. Let me assert these values at the outset and, instead of repeated and boring clarifications, offer only one more self-conscious explanation. In the sociology department of my university, the term for the 'homelands' is bantustans – which was employed during an earlier period in the evolution of apartheid ideology – and, although this sometimes leaves newcomers to the debate wondering what is being discussed, the term is taken to show a firm and unyielding opposition to apartheid. The use of 'homeland' is viewed as a bit wishy-washy, perhaps because it is the most commonly used term and is employed by such liberal institutions as the South African Institute of Race Relations. On the other side, inside government institutions, the reference is to 'self-governing and independent national states', which clearly they are not. I have persisted in referring to 'homelands' without intending to foster notions of tribalism and without suggesting that the relevant homeland somehow represents the appropriate area to which the ethnic group in question can make claim.

The first draft of this book was written in 1986/7 while I was a Visiting Scholar in the Department of Urban Studies and Planning at the Massachusetts Institute of Technology. My sojourn there enabled me to obtain the comparative international material that is central to the book. I am therefore especially grateful to Professor Lloyd Rodwin, who encouraged this enterprise and organized my visit, and to the Anglo American and De Beers Chairman's Fund which provided the financial support.

Many people commented on all, or part of, earlier drafts and offered suggestions that improved the clarity of my arguments. Colleen Butcher, David Cooper, William Doebele, Tom Karis, Adrienne La Grange, Mike Morkel, Pauline Morris, Lisa Peattie, Ivor Sarakinsky, Charles Simkins, Alfred Van Huyck, Nick Vink, Dan Weiner and two referees employed by Unwin Hyman are worth special mention. Some of these names are fairly illustrious and I was at first chagrined when they took it upon themselves to do more than just decorate a preface. On reflection, I found their criticisms insightful and they caused me to dig deeper for unobstructed points and consistent arguments.

A common carp was that my writing was turgid. One might claim good company, for the same could be said of Marx but, given my goal of furthering debate among a wide audience in the area of urbanization policy, I eventually turned to Dee Wingate for help with this aspect of the book.

My wife, Mary, interrupted her career to accompany me to Cambridge, and she and my two children, Catherine and Lisa, put up with an absorbed and too-often unavailable husband and father. Thank you.

URBANIZATION
IN POST-APARTHEID
SOUTH AFRICA

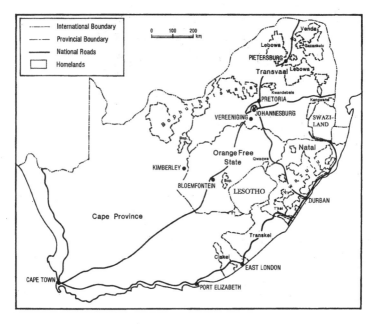

South Africa: provinces, homelands and cities

1 Introduction

My primary aim in writing this book is to promote debate about the urbanization policies appropriate to a democratic South Africa. In attempting this, I address three prominent urbanization issues:

- the housing shortage, accentuated by rapid urban growth;
- the desirability and feasibility of diverting economic and population growth from the Pretoria–Witwatersrand–Vereeniging (PWV) region; and
- the fostering of development in the rural areas in order to reduce pressures towards rural–urban migration.

The last issue includes an examination of the effects of redistributing white farms and an assessment of how to respond to the growth of towns in the 'homelands'.

This enterprise was conceived in 1985 when the intensity of the liberation struggle caused many to begin to question more deeply the nature of the post-apartheid future.[1] However, the imminence of majority rule, so strongly felt then, has since receded – there is now widespread acceptance that unless the apartheid government decides that it is advantageous to negotiate, it cannot be dislodged soon. Yet, in a contradictory fashion, the restrictions imposed early in 1988 on the United Democratic Front (UDF) and sixteen other progressive organizations have further accentuated the prominence of the African National Congress (ANC) as the leader of the struggle. It is ever more likely that a democratic South Africa would usher the ANC into power.

The strength of the National Party (NP) and the more certain position of the ANC as the heir-apparent lead the book in different directions. On the one hand, it would be useful to be able to begin to identify policies that, prior to majority rule, might form the substance of what Adam and Moodley (1986) term 'real reform'. On the other hand, there is still considerable value in debating the urbanization policies that an ANC-led government might consider. Some resolution of this divergence is found in the fact that the issues surrounding urbanization will largely remain the same, regardless of whether South Africa is governed by the NP or led by the ANC.

For example, there is already an intractable problem regarding the ability of the urban poor to afford housing and this will not simply disappear when the ANC comes to power.

Differences between the NP and the ANC are more predictable in the area of policy responses but, even here, one can argue that, when the NP considers 'genuine reform' (Yudelman 1987), the differences between it and the ANC will narrow. I am alert to the fact that this position might appear surprising to many people and perhaps even insulting to supporters of the ANC. I hope the following will clarify what is being asserted.

The reason for being interested in establishing a measure of genuine reform is found in the likelihood that the NP will continue in power for some years to come, and also in the state's ever increasing allocation of resources to the urban sector. Hitherto, progressive forces in South Africa scorned the government's 'authoritarian reform' because the 'reforms' were not a product of negotiation. More recently, the evident strength of the state, in contrast to the apparent weakness of the opposition, has led many to re-evaluate this position and to conclude that, if the end to apartheid is still some way off, then genuine reform warrants less disdain. If this book helps one to distinguish between such reform and other policy shifts, then decisions regarding which policies to support and which to oppose should become all the more apparent.

An illustration of the value of having a measure of reform is found in housing policy: at the moment the government is urging the private sector to take primary responsibility for solving the housing crisis, whereas the Freedom Charter raises images of public housing. Neither response will address the needs of the poor. On the one hand, the housing needs of the poor do not constitute effective demand and the (formal) private sector will not help them. On the other hand, public housing is also of limited value since its cost has prevented countries, both capitalist and socialist, from building enough dwelling units. Public housing has a particularly dreary record internationally in that it has invariably been used to house bureaucrats and people having, say, incomes falling between the 30th and 50th percentiles.[2] So the NP and the ANC may differ regarding the form of a desirable housing policy, but both should be disciplined by what has proven effective elsewhere. The consideration, not only of what is desirable, but also of what has 'worked' considerably narrows down the options – one finds that seemingly opposed positions are selecting policies from much the same menu.

While there is a tendency to reject government policy simply because of its apartheid origins, the above conclusion causes one to dissect that policy more closely. One should accept that, although most aspects to the government's reform agenda deserve rejection, some may be worth supporting. The important point is that one should be able to make these distinctions and that one should make them with a view to a democratic future.

To sum up, this book is intended to be used both as a target and as a weather-vane. If it prompts discussion regarding the post-apartheid future, the reason for writing it will in large part have been realized. If, in addition, the policies can be taken as a measure of the distributional decency of the government's urbanization policy, then the reader will be able to distinguish more clearly between the rhetoric and the substance of reform.

A focus on poverty

Most people think of reform in South Africa in terms of a movement towards democracy, but this is not entirely true for urbanization policy. The reason for this is that if 'you abolished apartheid most South Africans would remain poor and lead lives of material deprivation' (Freund 1986a: 124). The desirability of alternative policies is therefore primarily reflected in the extent to which they serve the material needs of the poor, and it is in this regard that I have asked, 'Which policy best contributes to the alleviation of poverty?' The assumption underlying this focus is that policies that relieve poverty arise only as a result of pressure from the poor. I am attempting to draw forth effective options with the hope that this will enrich the political process when policies are debated.

It is a bit disconcerting that a focus on poverty leads to an examination of the urbanization issues that affect blacks. Ordinarily, I would want to go along with the use, among progressive organizations, of the term 'black' to include 'coloureds'[3] and Asians, but it will become apparent that the material circumstances of the different groups differ so markedly that the narrower use of the term is more accurate when discussing urbanization policy. This contradicts the aracial future that one hopes for in a post-apartheid society and consequently requires further explanation. It is clear from Table 1.1 that future urban growth will preponderantly comprise the urbanization of a low-income black population – they are the poorest group in the country, constitute about 75 per cent of the population, have the highest population growth rate and an urbanization level that is well below that of the other groups. A number of specific problems follow: perhaps more than half the black population cannot afford cheap houses, which are priced around R20,000 (including land); areas zoned for whites, coloureds and Asians have a housing shortage of about 55,000 units, whereas in 'black' areas in the PWV alone there is thought to be a housing shortage of 350,000 units; it is blacks who suffer most

Table 1.1 South Africa: population size, growth rate, urbanization level, racial proportions and income distribution

	Asian	Black	Coloured	White	Total
Population total and exponential growth rate					
1990 36,481,000	1.58	2.59	1.86	1.53	2.32
2010 54,230,000	1.02	2.00	1.22	1.11	1.80
Proportion of total population					
1990	2.6	74.8	8.5	14.3	
2010	2.2	77.8	7.6	12.4	
Urbanization level, 1990	90.6	50.0?	80.6	89.4	
Income group per month before tax in 1985 (R)					
1– 99	0.3	22.5	7.2	0.2	
100– 199	2.5	16.8	13.4	1.1	
200– 299	5.4	17.1	10.5	1.1	
300– 399	6.4	10.2	7.9	1.6	
400– 499	7.8	9.6	7.9	2.2	
500– 599	8.0	7.3	7.1	2.4	
600– 699	6.7	4.8	6.5	2.1	
700– 799	7.3	3.7	7.1	2.8	
800– 899	6.9	2.2	6.6	3.4	
900– 999	5.5	1.6	4.8	3.1	
1,000–1,999	31.1	3.5	17.9	37.0	
2,000+	11.9	0.7	3.1	43.0	

Sources: Calculated from Johnson and Campbell (1982), Simkins (1983) and De Vos (1987).

from the inequitable access to land; and so on. Accordingly, it is not out of choice or any desire to speak in racial terms, but because blacks are the most poverty-stricken group, constitute such a large majority and are little urbanized that this book has mainly to do with the urbanization of the black population.

Yet, while the above simplifies the problems that arise when aggregating South Africa's racially differentiated data, it does not allow easy generalizations regarding the nature of poverty. Poverty is an elusive concept and I would be foolish should I claim a prior understanding of the problems of the poor. There is even considerable debate as to what poverty means! I therefore conclude this chapter with a review of the concept and a description of both the current and likely extent of poverty in South Africa.

Current urbanization policy is largely motivated by the security needs of the state and a short description of the policy and the negative consequences for black welfare is contained in Chapter 2. In the words of one commentator on this text, the chapter looks 'at the apartheid system as the troubled soil into which the new government must sink its foundations',[4] as it shows that the legacy of apartheid will long burden future governments and the economy. This burden is particularly apparent in the structure of the 'apartheid city',[5] and an extended example demonstrates the relationship between urbanization policy and poverty, and points to the nature of the problems a democratic government would encounter.

A bastardized city structure has arisen under apartheid. Pretoria is an extreme case, but Johannesburg also has a discontinuous urban structure, with pockets of isolated black residential development located at great distances from urban centres. This city structure is in direct contrast to the needs both of the poor and of an efficiently functioning city. Deprived communities in a city are dependent on 'access to the economic opportunities and the social and commercial services which can be generated through the agglomeration of large numbers of people' (Dewar 1986: 1). Access to employment and markets for the informal sector would be enhanced if urban activities were spatially integrated and occurred at high densities. Such conditions would also make an effective public transport system possible. In contrast, the black townships, including those currently being created, reflect quite the opposite characteristics and they represent the isolation of the poor from the relatively wealthy. The consequence is a paucity of social and commercial services and an expensive transportation system.

This urban structure has been a deliberate artefact of apartheid policy, whose primary purpose has been the attempt to keep blacks at a 'safe' distance from the centres of production and the white residential suburbs, except, of course, for labour purposes. South Africa's riot-torn cities are remarkable for the fact that most whites live and work 'inside the laager' which means that they seldom feel endangered, let alone materially affected, by the struggle – the riots occur largely in the townships. In effect, the access of the poor to employment and services, as well as the quality of their environment, are deliberately sacrificed in the interests of an illegitimate government's attempt to maintain social control. A democratic South Africa would obviously attempt to redress this situation of urban isolation.

Policy options?

It is not too much of a caricature to claim that much progressive analysis of the post-apartheid future criticizes the current system at

length, briefly considers the Freedom Charter, and concludes that
'the people will decide'. As Gumede, co-president of the UDF,
suggests below, it is surely possible to render a greater service:

> Paul Bell: It seems that in any discussion of a new society
> by the progressive movements, more time is devoted to a
> critique of the present system than to the presentation of
> alternatives. . .
> Archie Gumede: I agree. This is a situation which I cannot say
> is avoidable in our particular circumstances. We are on the run,
> and trying to be creative at the same time. I honestly become
> very worried, though, that sooner or later we are going to find
> ourselves in a real jam, because people do not have sufficient
> time to think, to discuss the various issues and possibilities.
> (*Leadership*, 1987, 6: 54)

All this is not to argue that policy can be satisfactorily considered independently of its context, and a hypothesized political
–economic context is discussed at length in the third chapter.
On the assumption that the topic holds intrinsic interest for the
reader, more attention is accorded to the competing images of the
post-apartheid state than one would normally expect in a book on
urbanization policy.

To start with, there are those who support a federal state and
capitalism. This position is represented, in one form or another,
by the 'left' wing of the NP, the opposition Democratic Party
(DP), whose views largely represent an extension of the former Progressive Federal Party (PFP), and major business interests. Various homeland leaders, most respectably the late Phatudi
and Buthelezi, also support federalism, but their opinions are
likely to prove of little account.[6] Next, there are those who
support a unitary[7] state and economic arrangements that range
from a mixed economy to socialism. It is the formidable alliance
represented by the ANC, the Congress of South African Trade
Unions (Cosatu), the UDF and the more recently formed Mass
Democratic Movement (MDM) which is examined here. (The
MDM comprises a loose alliance between Cosatu and the UDF,
and other organizations, has no constitution and office bearers, and
is a broad-based movement whose structure is intended to enable
it to evade repression.) While there appears to be unanimity on
the goal of a unitary state, it is in the nature of the ANC as
an opposition alliance that there is considerable debate within
the organization and its allies regarding the desired economic
arrangements. This debate is examined through reference to the
Freedom Charter,[8] the ANC's recent constitutional 'update' of the

Charter, views within Cosatu, and the ANC's relationship with the South African Communist party (SACP).

The discussion of the federal and capitalist image and the unitary and mixed economy image sets up two poles from within which a negotiated post-apartheid state is likely to issue. An important premise underlying this book is that the post-apartheid state will in fact be a product of negotiations. From the point of view of predicting the nature of the future South Africa, much would appear to depend on the manner in which the next government comes to power: if it occurs as a result of negotiation among relative equals, some compromise among the above options is likely; if it is the outcome of a victory in a civil war, then the extent of socialism and centralization and the absence of democracy is likely to be pronounced.

On the basis of presumed negotiations it is hypothesized that the post-apartheid state will have a unitary constitution and a socialist visage but will in practice constitute a mixed economy. In other words, the suggestion is one of a concession to the ANC regarding the constitution, but that business and Western interests, together with African nationalist elements within the ANC, will prevail in respect of the economic arrangements. Should this hypothesis prove correct it is apparent that, under a reform-minded NP and the ANC, there is still greater reason than hitherto imagined for urbanization policy to coalesce. Both operate with a centralized government and a mixed economy, with the differences between the two largely centring on the determination of the ANC to ensure a more reasonable distribution of the country's wealth.

In addition to the misgivings some people have regarding the possibility of considering policy options in advance of a clearer political–economic context, others have questioned the usefulness of the ensuing debate owing to the absence of reasonable data. The latter have maintained that, while the present empirical picture on social issues is unclear, in the future it will be decidedly muddy.

There are two possible responses to this uncertainty. One is that South Africans have seldom had reliable urbanization data. To illustrate: recent official estimates of Soweto's population have been 700,000, 800,000 and 1.5 million, while unofficial estimates point to a figure of 2.5–3 million. Government acknowledgement of the latter figure would represent an acceptance of the extent to which its 'influx control' policy had failed. The result of such uncertainty is that it has never been possible to do more than hope one has a good assessment of underlying trends and choose a demographer whose assumptions seem the most reasonable.

The second response is that empirical clarity is possible in very few countries and there is little reason to expect much improvement in the future. The standards expected of South African data should reflect the country's level of development. I argue below that South Africa is increasingly typical of middle-income countries and, in this light, it would be an optimist who assumed that the situation would improve. One works with projections and scenarios, and adjusts them as more data become available and as events unfold. But this is not to despair entirely of reasonable estimates – the workers who will be seeking jobs 20 years into the future are already alive; the same applies to those who will be forming new households and seeking land and housing. The general trends are already fairly evident. One is confident, for example, that the Eastern Cape will continue to be a lagging region and that the PWV, the country's economic core, will continue to display the problems associated with rapid urban growth. In short, there is already a reasonable picture regarding major urbanization problems.

In my view it is certainly possible to debate policy options constructively, and to fuel this debate I have based this book on the juxtaposition of existing and anticipated urbanization problems in South Africa with related problems and policies elsewhere in the world. I have usually compared South Africa with other less developed and middle-income countries, but I have also looked briefly at Eastern European countries in order to ascertain whether the consequence of not hypothesizing a socialist South Africa economy has been to forgo socialist solutions. It will be seen that there is considerable international experience with the types of urbanization problems that South Africa confronts. Recourse to experience and the latest theory must surely improve one's understanding of the issues and of potential responses to them. My objective is a clearer understanding, not only of what is desirable, but also of what is possible.

Urbanization policies

The actual discussion of the issues and options is based on case studies, which deal with:

● land for housing the poor;
● regional policy and attempts to divert population and growth from the PWV; and
● black occupation of subdivided white farms and urbanization pressures in the Transkei.

Under NP rule (and before) the primary urbanization 'problem' was perceived to be black migration to the cities and policy was accordingly directed at restricting or diverting urban growth. The foremost urbanization problem under a democratic government will be how to accommodate low-income urbanization. For this reason, the fourth chapter considers land for housing the poor. I have used the case of the Witwatersrand[9] because it is here, at the country's core, where urban growth will be most rapid and, consequently, where any government's housing policies will be most sorely tested. Simkins (1985) estimates that between 1985 and 2000 the PWV's black population will grow by about 5 million people and approximately two-thirds of the increase will locate in the Witwatersrand.

The history of Third World housing policy is assessed – public housing briefly, housing policy in socialist countries equally briefly, and site-and-service schemes and public–private partnerships in greater detail. No solution is seen in any of these policies for South Africa's poor. Instead, if their needs are to be met, public policy directed at increasing the supply of well-located affordable land is held to be central. International examples of how to pursue such a policy are considered and recommendations are made for South Africa's urban areas.

I examine regional economic policy in the fifth chapter. The case study material is based largely on the growth of the PWV relative to the nation's other cities. Given the notoriety of 'regional' policy under apartheid, it is ironic that there should be a chapter on regional policy, but it seems likely that a democratic government would persist with policies directed at diverting growth from the PWV. Governments attempt to control the growth of large cities throughout the world and a future democratic government of South Africa is unlikely to be any different.

Despite all this, it is not clear that the PWV's growth should be diverted or whether regional policy would be able to do so. Regional planning is a discipline in considerable disarray and the ability of public policy both to direct the location of employment and not to hamper the efficiency of the economy is unproven. My expectation is that ethnic/regional competition for limited resources will press the need for regional policy and, also, that there might be pressures within the PWV itself to resist further population growth. For instance, if one considers that an ANC-led government's primary constituencies are petit bourgeois groups and organized labour, it is precisely these groups that are in competition with migrants for land and housing, municipal services and jobs. Elsewhere in the world it is frequently the case that such groups support restrictions on urbanization.

The attention that regional policy will receive is matched by its lack of a sound motivation. In comparative terms, the centres comprising the PWV are small and, ultimately, there is reason to believe that the issue is not spatial. In the light of the existing and anticipated levels of poverty in the country, the point is not so much to create biases for or against the PWV but to remove distortions that harm the efficient operation and growth of the economy. The issue would more beneficially be phrased as 'Would the unrestricted growth of the PWV enhance the country's economic development and would it facilitate employment creation, in both the formal and the informal sectors?' Chapter 5 consists, in part, of the suggestion that a democratic government should resist the pressures for attempting to direct the location of economic growth. Little hope, though, is held out for a sense of proportion on the topic and there is likely to be frustration with planners owing to the paucity of effective policy recommendations.

In the hope of identifying a few such policies, international experience, again including that of Eastern Europe, is reviewed. Secondary city strategies are found to be foremost among the regional planner's 'tools of the trade'. Rondinelli (1983) holds that such cities offer the most potential for the creation of alternative foci which might divert migrants from heading for the largest city. However, such strategies have had little effect elsewhere in the world. Thus, while a secondary city strategy is suggested for South Africa, it would be inadvisable to expect much from it and it is offered as the least inefficient policy option.

Urbanization in the homelands and the redistribution of white farms are considered in the sixth chapter. Although urbanization policy usually goes hand-in-hand with rural development strategies, there is no evidence that the strategies work in terms of restraining rural–urban migration. Rural development, with only few exceptions, is labour displacing. Nevertheless, owing to South Africa's specific circumstances, it is particularly interesting. An initial reason for looking at the rural side to the urbanization equation follows from the view that, since the black populations of small rural towns in 'white' areas have diminished enormously as a result of forced removals to the homelands, would they not be able to 're-absorb' population following the demise of apartheid? However, enquiries in the Eastern Cape in 1987 revealed unemployment levels in the towns of 40–50 per cent and I was rapidly disabused of this notion, at least in so far as the Eastern Cape is concerned.

A more substantial reason for examining rural development is that the redistribution of farming land presently owned by whites would seem to be a marvellous anti-unemployment, anti-urbanization and regional development strategy. Given the experience of Kenya and many other countries, is there reason to advocate such a strategy? Is it really the case that, for once, efficiency and equity gains coincide? The question will be examined through reference to four magisterial districts in the 'white corridor' between the Ciskei and the Transkei. This entire region is an especially impoverished area of the country. One homeland, the Ciskei, has been the destination of considerable population removals and is especially dependent on migrant remittances and commuter incomes; the other homeland, the Transkei, is also impoverished but is relatively better endowed with agricultural resources. The corridor contains a strip of 'white' farming land and a number of small towns and it is here that, under a democratic government, one should anticipate strong pressures for land redistribution.

The last section of Chapter 6 contains an examination of urbanization processes and pressures in the homelands, using the Transkei for illustrative material. An attempt is made to determine how the towns can best serve their own and their districts' populations with respect to employment creation and income generation and, also, in respect of the demand for and supply of commercial and social services. As usual, it will be seen that there are few obvious answers! For example, urban growth in the Transkei is presently largely occurring in peri-urban areas outside the larger towns and also in some rural villages. This form of urbanization is a response to the high cost of land and services in the towns, but it brings with it a number of problems. Growth, for instance, is located within areas controlled by chiefs who have neither the ability nor the finances to organize the provision of the necessary services.

Moreover, is it correct to postulate into the future on the basis of current trends? What proportion of migrant labour from the Transkei would prefer to settle in the cities? Would 'Transkeians' accept a progressive loss of land as a 'master farmer' class emerged? Given unemployment throughout the rest of South Africa, would the rural and peri-urban unemployed move to the larger cities? This chapter was particularly challenging as it involved speculating in a most uncertain world.

For students of urbanization, the above problems must appear remarkably incomplete since they exclude a concern with local government. There is now considerable consensus that policies intended to restrict urbanization do not work and frequently involve profound infringements of individual liberty. There is also agreement that policies intended to redirect urbanization to secondary cities, or to contain it through rural development policies, are either unsuccessful or have only marginal effects. The result has been a redefinition of the problem, with the focus shifting from restricting the growth of core cities to the better management of that growth, and it is in this light that local government is considered to be a central urbanization problem in Third World countries.

South Africa, however, is in some respects decidedly not Third
World. It has developed financial markets, transport and communication systems. It also has a long tradition of effective local
government and the administration of urban affairs, albeit frequently employed for repressive purposes. The local government
problem arises from the separation of wealthy and poor parts
of town, with the poor parts having historically been denied
industrial and commercial development and thereby a stronger
financial base, and having had illegitimate local authorities imposed
on them. The country currently bears the burden of these 'apartheid
institutions', but, with majority rule, the integration of black and
white municipalities, and the sharing of administrative capacity
and the (still too limited) finances of the rich parts of town, the
local government and planning problem will greatly diminish.
In essence, the issue is the realization of majority rule and the
constitutional outcomes that follow from it. So local government
is not viewed as an urbanization problem for the purposes of
this study.

A last consideration has to do with the tone of the analysis and
the ordering of the material in the case studies. The intended
tone is revealed in the book's purpose. If the book degenerated
into a prescriptive treatise, the effect, as Wolpe[10] has warned,
would be that many readers would be 'turned off'. His view,
which I share, is that the book's impact will be greater if it
achieves a measured elucidation of alternatives. I hope that an
argumentative and enthusiastic personality has not caused me to
contravene Wolpe's injunction too often! As regards the material I
have used, it is intended to be accessible to any person wishing to
become informed on these issues. This goal means I have attempted
both to look at the latest thinking on a topic and also to explain
how it is one got to the present impasse. The policies are therefore
examined in an international and historical context: what policies
were employed elsewhere, why did they succeed or fail, and what
is their relevance for South Africa?

A certain naivete?

In a different context, Wellings and Black (1987: 182) have pointed
out that 'whilst planners might wish to see what could be done
within the existing political framework to improve conditions in
the homelands, they should also consider to what extent their
recommendations are likely to strengthen those structures which
exclude the poor from access to material resources and political
influence'. How can this text escape doing the government a similar
service?

I have two possible responses. First, my focus is not on how the government might retain white supremacy but on how public policy might alleviate poverty. If reforms are suggested that would improve the material conditions of the poor but that do not at the same time threaten the state, this is not to be criticized. It would be incorrect to believe that the further impoverishment of blacks will hasten the revolution and that their poverty represents an inevitable, even necessary, short-term cost. We are not witnessing the last stages of the struggle. In this context,

the minimization of suffering here and now seems to us a worthy goal even though it may occur at the expense of a more noble dream; to postpone small-scale reform in the hope that misery will accelerate a more fundamental transformation to us smacks not only of cynicism but of immorality. Indeed, it is true that apartheid cannot be reformed but must be eradicated. Yet this dismantling of a political system does not necessarily require the destruction of a society. It is an illusion that an alternative can only emerge from the ashes. If this were so, it would hardly be worth the price. At the same time, real reform must be rescued from its present proponents in Pretoria. The demystification of reform rhetoric must be high on the agenda of anyone interested in speedy evolution.

(Adam and Moodley 1986: 8)

Secondly, many of the policies to be discussed presume racial redistribution of public spending and democratic participation in decision-making. An effective redistribution of land, for example, cannot occur without the participation of those affected. In other words, many of the options identified are available only to a government that is either democratic or undertaking genuine reforms.

Poverty in South Africa

The purpose of this section is twofold. One is to introduce the concept of poverty and to emphasize, in particular, that poverty results from circumstances that have little to do with the poor. Deprivation in South Africa will be seen to result from both the relative powerlessness of the poor and the fact that South Africa is increasingly a labour-surplus economy. The second purpose is to introduce summary statistics regarding the likely extent of poverty in the country.

It is important, at the outset, to specify that two notions of poverty – relative inequality and absolute poverty – are usually implicit in most discussions of this topic. A cynic once noted that Marxists focused on absolute poverty during the Industrial Revolution but, once such poverty ceased to be a problem, they were left without a cause.[11] The result was a rapid change of horses and a focus on relative inequality. There is considerable truth in this view, in so far as the priority one accords issues of absolute poverty and relative inequality does seem to shift as a country becomes more developed. South Africa, of course, will prompt considerable debate about the appropriate focus, precisely because it is a middle-income country. However, the view taken hereafter emphasizes absolute poverty. This is because the hypothesized context of the book is that of a democratic society within which the political causes of relative inequality will be less pervasive and because unemployment will have increased.

An empirical assessment

In the case of relative inequality, it is frequently held that until the 1970s South Africa had the most unequal distribution of income for countries for which the data are available. The Gini coefficient of employed persons in 1971 was 0.71 (Simkins 1979), which denotes a remarkably unequal distribution, but by 1980 the coefficient had declined to 0.57 (Devereux 1983).[12] The latter statistic compares with other middle-income countries – it is better than in Brazil and slightly worse than in Mexico. Thus Knight (1986) could say that by 1980 apartheid had been more guilty of structuring inequality along racial lines than of producing that inequality; and Blumenfeld (1986: 2) could compare South Africa with other countries that, 'though not subject to the distortions and constraints imposed by apartheid, are also characterised by endemic poverty, illiteracy, inequality and unemployment.'

The problem with these data is that, in an environment of increasing unemployment, they have less and less value in describing overall levels of inequality – that is, including unemployed persons and their dependants. The contrasting Gini coefficient of all households in 1980 was 0.66 (Wilson and Ramphele 1989). It follows that since:

- the rapid decline in inequality occurred alongside increasing employment opportunities and growing incomes;
- the future scenario is one of rapidly increasing unemployment (see Table 1.2);
- levels of inequality typically decline as a country becomes more developed;

the certain means of reducing inequality and poverty is economic growth from the ranks of middle-income countries. The material circumstances of the people require rapid growth.

The high level of inequality also points to markedly unequal racial shares of the national income – the extent is evident if one compares population size with income share in Table 1.1. In 1970 the white share of the national income was 71 per cent. This share declined to 61.5 per cent in 1980 and is expected to decline further to about 54 per cent by 2000 (Van der Berg 1986). When contrasted with the white percentage of the total population of 15.5 per cent in 1980 and 13.3 per cent in 2000, the extent of the racial inequality is fully revealed. In addition, the decline in the white share of the national income is overstated owing to the increase in the black share of the total population.

The improvement in the 1970s reflects changes that occurred during the first half of that decade, namely better black incomes coupled with rapid increases in employment. In the later 1970s, and ever since, reductions in the racial wage gap have continued. For example, between 1972 and 1984, the total real wage per black employee increased by 6.6 per cent per annum, whereas that of whites increased by only 0.6 per cent per annum (Knight 1986: 16). However, the improvement has been accompanied by higher black unemployment, which grew from 11.8 per cent in 1970 to 21.1 per cent in 1981 (Simkins 1982). In other words, the growth of the black share of the national income should not automatically be taken to represent a general improvement in the black standard of living. Indeed, Devereux (1983) draws the unhappy conclusion that the improved position of the black working class may well be 'at the expense of the Black masses'.

Future prospects for black employment are even more depressing. World Bank estimates (cited by Knight 1986: 13) have it that the black labour force grew by 2.9 per cent per annum between 1970 and 1982, but, due to demographic lags, will grow by 3.3 per cent per annum between 1980 and 2000. However, employment growth in the formal sector declined from 2.5 per cent per annum in 1946–75, to 1.4 per cent in 1976–82, and to 0 per cent in 1982–6 (Spies, n.d.).

The implications for black unemployment are seen in Table 1.2. (The peripheral sector is taken to include 'unemployment, subsistence agriculture, the informal sector and what could be called "non-employment".) It illustrates that formal black employment is due for a precipitous decline to 45.1 per cent of the labour force in the year 2000 and to continue to decline thereafter. This forecast is not exceptional; for instance, Van der Berg's (1986) estimate is a bit less depressing at 48.5 per cent (assuming a 3.2 per cent per

Table 1.2 Distribution of the black labour force between the 'formal' and 'peripheral' economies

	1980	1990	2000	2010
Formal	61.5	52.8	45.1	39.3
Peripheral	38.5	47.2	54.9	60.7

Source: Roukens De Lange (1984).

annum growth between 1980 and 2000), whereas Spies' (n.d.: 7) prediction, based on the growth pattern of the last decade, is much worse. He writes: 'only 33% of the Black labour force will be able to find employment . . . by the year 2000.' If one puts the slow growth and increasing capital intensity of the South African economy together with the rapid growth of the black labour force, then the predicted unemployment levels are seen to reflect long-term structural causes.

It is also worth noting that, for most people, the informal sector does not provide a substitute. This is because it largely serves the consumption needs of the low-income labour force in the formal sector and, as the proportion of formal sector opportunities decline, so do those in the informal sector. Moreover, labour in the informal sector involves lower wages and worse working conditions. Since 'the relationship between unemployment and poverty is immediate and strong' (Simkins 1986: 9), and since the informal sector is a poor second best, one cannot expect any reduction of the extent of poverty in South Africa.

At this point it is instructive to consider the following summary of the findings of the Second Carnegie Inquiry into Poverty and Development in Southern Africa of 1984:

In 1980 about 12 million South Africans had incomes below the poverty line. Of these, 9 million lived in the bantustans. Their households were on average R120 short each month of the poverty line income of R200. A full 1.4 million people were without any measurable income.

A further 2.2 million impoverished people were to be found on the farms of 'white' South Africa – that is, one in two farm dwellers. In the townships, 1.3 million people were to be found eking out sub-poverty-line existences.

In the homelands most families were – and are – dependent on remittances of migrant workers, which averaged at a quarter of the average monthly pay of R189 in 1980.

(*Star*, 4 April 1984)

In the light of the predicted peripheral location of future black employment, it is difficult to foresee a decline in the extent of poverty in South Africa. Even if one assumes, for example, that the post-apartheid period will see improvements in the circumstances of organized labour relative to capital, there remains little basis on which to hope for improvements in the conditions specified above, as it is less and less the case that the poor are members of organized labour. It is because of the declining relative inequality in racial wage differentials and the rapidly increasing proportion of the labour force that is un- or underemployed, that absolute poverty appears to be the proper focus of urbanization policy.

The concept of poverty

It follows from the above discussion that the poor are not poor through some fault of their own. It is, I hope, by now a commonplace that the poor are not materially deprived because they are inept, lazy or, in some volitional sense, unwilling to undertake the sacrifices necessary for success. The tendency in the past has been to 'blame the victim' in this manner. Probably the best known such explanation has been founded on loose adaptations of Oscar Lewis' (1961) more subtle notion of a 'culture of poverty'. This notion postulates the rise of negative, self-defeating attitudes in response to a situation of deprivation. These attitudes – suspicion, apathy, cynicism, helplessness, inferiority – are further held to be passed on to subsequent generations, even when material circumstances change. The outcome is a vicious circle of poverty wherein the onus is placed with the poor. However, what one has here is a description of symptoms more than an understanding of causes. The marked initiative and competence displayed by squatter movements, for example, has given the lie to elite preconceptions regarding the attitudes of the poor. Perlman (1976) has shown that squatter attitudes are finely attuned to the opportunities available to them and realistically reflect those attitudes. Moreover, she notes that, to the extent that the urban poor are unrealistic, it is in their being over-optimistic (particularly the men).

The kinds of features that are typical of the poor are: being female or being a member of a female-headed household, youth, a lack of assets, disability, large families relative to the number employed, casual-labour status, ethnic affiliation, and recent immigration to the urban informal sector (Chambers 1980; Lipton 1985; Moser and Levy 1986). The ultra-poor (below the bread-line) are associated with a 'lack of human and physical assets; with weak labour-market positions; with large families, high dependency ratios, and very high infant mortality; and with significant risks of nutritional damage' (Lipton 1983: 3). Unemployment is also usually a critical factor, but is included in neither of the above lists in order to emphasize the point that the relationship between poverty and unemployment is not exclusive – poverty may well be an outcome of underemployment and low wages. In other words, most poor have some sort of job but obtain inadequate incomes (Rao 1974) and it is this situation that one expects to be most typical of South Africa's poor.

Conclusion

In wealthy countries like America and Sweden, where income per capita is high, one can argue that pronounced poverty is a measure of the relative political power of the poor. This position is less tenable in middle- and low-income countries since, if the economy requires too few labourers, if the assets available are insufficient relative to the number of claimants, and if the fiscal capacity of the government is constrained, then, regardless of the location of political power, the government can hope to relieve only selected instances of poverty. It seems unarguable that, the more developed a country's productive capacity, the more it is the case that poverty is socially determined.

Would the advent of democracy enable some future South African government to undertake rapid improvements in the allocation of resources? In 1984 Simkins showed that if one assumed about a 1 per cent per annum real growth rate in per capita incomes between 1976 and 2000, and interpreted the consequences of this slow (but possibly optimistic) growth for the budgetary capacity of the government, then racial equality in social expenditure in 2000 would imply a per capita expenditure about equal to that of coloureds and Asians in 1975/6. In other words, blacks would not see the rise they desire and whites would see a considerable drop. It is clear that there can be no simplistic assumptions regarding subsidies and public hand-outs. A democratic government could hope to achieve results only at the margin of pervasive welfare problems. It is in this light that the urbanization policy options available to future governments should be examined.

Notes

1   Probably the best-known example of this is the conference on 'The Southern African Economy after Apartheid' held by the Centre for

Southern African Studies of the University of York, 29 September to 2 October 1986.

2 The estimate is provided by Mr Alfred Van Huyck who, while I was on sabbatical at the Massachusetts Institute of Technology, was president of Planning and Development Collaborative International (PADCO) in Washington DC and visiting professor at MIT.

3 The term 'coloured' refers to people of mixed race. It is widely rejected by many coloureds, and others, and the common designation 'black' is frequently employed. My reasons for employing the expression are contained in the text.

4 The eloquence is that of Professor Lisa Peattie.

5 The notion of the apartheid city is not new; for example, see Davies (1981), Western (1982), and Lemon (1987a, 1987b).

6 This is also the case for Buthelezi. Repeated opinion polls over time show that his support is waning, even in Natal. His power now does not even encompass his specific region and ethnic group.

7 Liberals in South Africa have objected that the term 'federal' does not imply the state is not unitary. While formally correct, the semantic nicety does not change what the various sides mean by federal and unitary.

8 The full text of the Freedom Charter is contained in Appendix 1.

9 The Witwatersrand comprises the magisterial districts of: Central Witwatersrand (Johannesburg, Randburg, Roodepoort, Germiston and Alberton); East Rand (Kempton Park, Benoni, Boksburg); Far East Rand (Brakpan, Springs, Nigel, Heidelberg); Far West Rand (Oberholzer); West Rand (Krugersdorp, Randfontein, Westonaria). In addition, the Vaal Triangle consists of the magisterial districts of Vereeniging, Vanderbijlpark and Sasolburg.

10 In conversation.

11 The observation is that of Michael Brown.

12 The Gini coefficient takes values ranging from 0 to 1. A perfectly equal income distribution has a value of 0; conversely the closer to 1 the more unequal the income distribution.

13 The table presumes the persistence of economic conditions characteristic of the period 1976–84. During this period, taken as a whole, growth was below 2 per cent per annum, with a brief improvement at the beginning of the decade caused by higher gold prices.

2 Urban 'reform' under apartheid

The South African government has in the last few years enacted two major reforms in the area of urbanization policy. One involved the repeal of the notorious influx control legislation under which, in the past few years, about 200,000 blacks have been arrested annually for being in the cities illegally. The actual legislation was not so encompassing in that the touted reform was enacted for only about half the homeland population – those living in the 'non-independent' homelands. None the less, entrance visas or permits are not being required from the residents of the 'independent' homelands and, in effect, all blacks are now allowed access to the cities.[1]

The second major reform was the passing of the Black Community Development Amendment Act of 1986, which introduced home-ownership into black residential areas and instituted procedures through which the supply of land for black housing could be increased. Whereas black housing on the PWV was previously owned and leased or rented by the government, a black housing market was now initiated and the private sector was encouraged to assume responsibility for black housing. The government now considers applications for 'development areas' (residential land zoned for blacks), and there has indeed been an increase in the supply of land available for black housing. The government has, at the same time, withdrawn from the supply of housing and has indicated that it would in future concentrate its efforts instead on the supply of land and bulk infrastructure. One may deliberate the desirability of the government's attitude to its role vis-à-vis the supply of housing, but the obvious shortcoming of the Act is that it occurs within the framework of residential segregation (the Group Areas legislation).

Even though residential segregation and controls over the location and supply of land for housing blacks were retained, the government has presented these changes as major reforms. However, the artifice, mentioned above, of distinguishing between the two types of homelands, together with the residential controls, further alienates blacks and contradicts calls by liberal capital in favour of removing all racial biases to urban legislation. How, then, is one to understand government policy? My view is that the reforms largely represent no more than a change in the form of repression. Urbanization restrictions previously operated at the national level, limiting black urbanization and prompting migrant labour from distant homelands. With persistent black movement to sites within homelands on the metropolitan periphery – that is, sites where low-cost land was available that was not zoned for whites – and the economy's increasing need for a skilled and stable labour force, the migrant form of labour supply is giving way to commuter labour. This shift, which is essentially a change in the form and location of the labour supply, has led to the government undertaking policy adaptations. Influx control, in a different guise, is now being implemented at the level of the metropolitan region.

What the government is doing is deliberately accentuating the cost of land close to the cities; although it can be argued that land prices ordinarily decline with distance from the city, in this case the situation is greatly exaggerated. Black townships, especially informal settlements, are being located at sites that are unnecessarily distant from places of work. At the same time, rigorous controls have been instituted to prevent squatting on land not approved for this purpose. Access to convenient residential sites will be possible only for those who can afford the related land and service costs and, whereas previously rationing of housing and restrictions on the right to be in urban areas kept blacks out, now cost and affordability are being pushed to centre-stage. The blacks living in urban areas will be those with jobs and relatively high skills. The less well-off will be pushed out to the urban periphery, and the unemployed, who are unable to afford land, will either rent rooms or live in shacks and be subject to enormous overcrowding, or be denied urban residence – for them, even the government's site and service schemes will be too expensive.

The metropolitan structure that arises as a result of these policies serves the security needs of an illegitimate government. The view is that 'unemployment cannot be solved economically but only contained politically' (Saul 1986: 84), and the result is a thoroughly contradictory set of reforms. On the one hand, the economy's need for urbanized and skilled black labour, coupled with the reality of black population shifts, has led to the removal of influx controls. On the other hand, security concerns have led to the retention of a deconcentrated[2] metropolitan structure – the apartheid city.

This chapter contains a short description of past and present urbanization trends. The history of the urbanization policies that gave rise to these trends is well rehearsed and there is only passing mention of them. The form of the current policies, and their rationale, are of much greater interest and it is these issues that I have considered at length. Only when one understands the government's policies and motivations can one distinguish between its typical heavy-handed policy adjustments and those policies that effectively serve the interests of the people.

In passing, however, one should note that there are divisions within the National Party – Pretoria's decision-makers do not have a Machiavellian clarity. The reason for the government's hesitant and alienating approach to policy change is that there are separate and competing policy strands within the NP (Van Heerden 1986) and between the various government ministries (Greenberg 1984). Traditionally the NP is divided into the 'verkrampte' (conservatives) and the 'verligte' (enlightened), but Van Heerden (1986) postulates the predominance of a central 'Old Guard' who persist with federal[3] constitutional visions wherein 'Nats will keep ultimate control' (p. 36). Thus, urban reform is not denied, but, so long as the NP insists on ultimate control, the changes will be 'erratic and eclectic as the regime oscillates between the desire to modernise and to keep its power base intact' (Slabbert and Welsh 1979: 5). For this reason, urban reforms, which are essentially attempts to co-opt sections of the black community, lag far behind the agenda of the majority of the population and are received with disdain.

Urbanization trends

Schlemmer (1985: 167) has observed that even government officials acknowledge that 'no-one knows the real size of the black urban population'. Indeed, research in South Africa has always to begin with the premise that, more so than in most countries, data are used for political purposes and are unreliable. Because South Africa maintains neither comprehensive nor accurate official urbanization data, we have to struggle to arrive at reasonable figures. In this quest I have, by and large, relied on Charles Simkins, a politically astute economist and demographer who is based at the South African Labour and Development Research Unit at the University of Cape Town. Although some of his essays are now rather dated – from the early 1980s – Mabin (1989: 2) has commented that they 'still constitute the most substantial yet available to outline the demographic history of African urbanization'.

Simkins' estimate of the current and future urbanization levels of South Africa's race groups is shown in Table 2.1. While estimates for the urbanization levels of Asians, coloureds and whites are unlikely to be questioned, the level for blacks is much more complicated. In proposing various refinements to a black urbanization estimate of 32.6 per cent, the complex and overlapping effects of apartheid on black urbanization will become apparent.

The first equivocation follows from the increasing prominence of a commuter, as opposed to a migrant, form of labour supply. Both these forms of labour supply refer predominantly to male labourers. The migrant labourer usually lived in single-sex hostels and hitherto had to return to a homeland every year to ensure that he forfeited his claims to permanent urban rights, which might otherwise have been sought under the relevant legislation. There were 1,395,000 migrant labourers in 1982 (National Manpower Commission, cited in South African Institute of Race Relations 1985: 258,9). Commuter labourers, in contrast, most commonly live just inside a homeland border and commute to their job daily or weekly. There were 773,000 such labourers in 1983 (idem).

Migrant labour is a common practice in Africa and elsewhere in the world, but in South Africa it has been enforced through the absence of a free labour market, by official allocation of labourers to specific jobs and by the requirement that the labourers' families stay behind, at 'home'. Further, permission to obtain work in the cities was dependent on employer demand and, in the absence of such demand, blacks were required to remain in areas where they commonly were unable to make a living. While it is no doubt the case that many migrants would prefer to keep their families in the homelands and, to an ever greater extent, on cheap land outside the homelands to which they have access but which may be some distance from the cities, the extent to which this would occur is unclear. Mabin (1989: 4) points to circumstances where 'entire households have frequently not migrated as a whole, and while a base has been maintained by some member in rural (more recently

Table 2.1 Urbanization rates by race, 1980–2000[1]

	Whites	Coloureds	Asians	Blacks	Total
1980	88.4	76.6	90.6	32.6	46.8
1990	89.4	80.6	90.6	39.4	52.5
2000	90.4	84.6	90.6	47.3	58.2

[1] The table excludes peri-urban areas.
Source: Simkins (1983: 143).

simply non-formally-urban) areas, other household members have moved to town for longer or shorter periods'. All the same, it is equally sure that many migrants would settle with their families in the cities, and the extent to which under-urbanization has resulted from such labour practices is speculated on below.

Commuter labour results from the migration process to the cities being blocked – formerly by influx control, now increasingly by land prices and overcrowded conditions in the townships. The outcome is that many blacks 'dam up' behind homeland borders where land is relatively cheap but lacks an agricultural base. A displaced 'peri-urban' form of urbanization has arisen – the migrants are urbanized but in impoverished, commuter zones within homeland borders. Map 2.1 shows the peri-urban development that arose outside the PWV in the adjacent Bophuthatswana and KwaNdebele homelands.

Elsewhere in the country, blacks in both informal and formal black townships (for example, KwaMashu and Mdantsane) were excluded from official estimates of urbanization by redrawing homeland borders, with the result that the people in these huge townships are not counted as part of the urban population of the cities of which they form a part. If one were to take this into account and to redraw Simkins' table, the figure for black urbanization would increase by 7 per cent to about 40 per cent in 1980, and would be 58 per cent instead of 47.3 per cent by the year 2000, depending on the future form of urban growth. The black urbanization level is currently estimated at about 50 per cent.

Table 2.2 shows that, over the period 1950–80, the proportion of the black population in cities, towns and 'white' rural areas declined alongside an increase in the homelands, with the decline in the rural areas being especially marked. This table is interesting since it reflects the operation of the two underlying forces impacting on the location of blacks.

Table 2.2   The distribution of the black population, 1950–80 (%)

	1950	1960	1970	1980
Metropolitan	16.8	20.2	18.2	18.7
Towns	8.7	9.1	8.9	8.1
Rural	34.9	32.1	24.5	20.5
Homelands	39.7	38.6	48.4	52.7
Total	8,660,400	11,506,900	15,468,100	20,972,300

Source: Calculated from Simkins (1983: 53–6).

Map 2.1   Urban development around the PWV

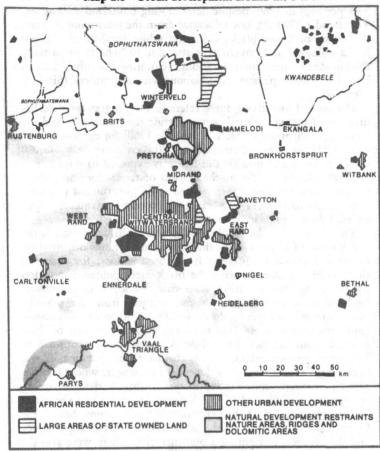

Source: Naude (1986)

One of these forces obviously consists of the influx control policies of the NP. The growth rate of the black population in the metropolitan areas exceeded that of the black population as a whole until 1960, but thereafter it was lower than the overall rate of natural increase. We can attribute this turn-about to an unrestrained government effort to realize the Verwoerdian home-land 'dream', which, during the 1960–70 period, gave rise to a massive coerced movement of blacks, primarily from the white rural areas into the homelands. Thus, whereas the proportion of blacks in the homelands dropped from 39.7 per cent in 1950 to 38.6

per cent in 1960, it increased markedly to 48.4 per cent in 1970 and 52.7 per cent in 1980, despite deteriorating homeland economies. The decade, 1970–80, saw relocation from the metropolitan areas, the move of fewer blacks from the white rural areas – about half a million – and government attention to 'wholly or partially closing down small town locations, relocating the inhabitants in homelands and replacing settled labour with commuters' (Simkins 1983: 62).

The second underlying force refers to the progressive mecchanization of the agricultural sector outside the homelands and the subsequent dismissal of black labour. In 1960, for example, there were 1,221,000 farm labourers; by 1980 there were only 973,000 such labourers and in 2000 the figure is expected to diminish to 780,000. Moreover, the mechanization of farms does not reflect the operation of market forces alone – the government provided concessions to farmers in advance of economic viability, which enabled them to replace labour with machinery.

Yet, Table 2.2, while probably the best available as regards the distribution of blacks outside the homelands, ignores urbanization in the homelands themselves. It has been shown, for example, that Simkins underrepresents the black metropolitan population by excluding the peri-urban areas that developed in the home-lands adjacent to the cities or that resulted from the periodic redrawing of homeland borders. In addition, other forms of home-land urbanization arose over the same period. The first of these resulted from the fact that, while 'Migrant labour continued to be numerically important, . . . its character was changing. It was becoming a workforce solely dependent on wages, whose families were increasingly concentrated in quasi-urban settlements in the countryside' (Hindson 1987a: 72). Secondly, there was the urban growth associated with the expansion of the economic base in the homelands, as a result of the industrial decentralization policy and the creation of homeland bureaucracies. There were also the population removals, which located blacks without an agricultural base in the homelands. While many gravitated to peri-urban areas on the fringe of South Africa's cities, some remained in or moved to urban settlements in the homelands.

Graaff (1986) suggests an effective urbanization rate in the home-lands in excess of 40 per cent. He is referring here to people dependent on incomes earned in urban areas and living in urban areas, in the peri-urban areas surrounding towns in the homelands (not just those adjacent to cities in the rest of South Africa) and in settlements in excess of 5,000 persons. Unfortunately, because this figure includes the peri-urbanization estimate of 7 per cent already noted with respect to Table 2.1, one cannot simply adjust

Simkin's earlier calculations so as to take Graaff's estimate into account.

Graaff's work is also problematic because the additional forms of homeland urbanization he identified make it difficult to conclude with a neat summary to the effect that apartheid has resulted in an under-urbanization of a specific percentage. It is apparent that there is a higher effective level of black urbanization than hitherto identified, and this leads to an interesting hypothesis: on the basis of an international comparison of other countries with South Africa's economic structure and level of development, Simkins suggests that the country's metropolitan population would increase by about 2.3 million persons,[4] and that this would almost exclusively comprise an exodus from the homelands. Were this to be the case, the level of black urbanization would be somewhat more than 10 per cent higher and the measure for the population as a whole would be about 8 per cent higher. If we now recall the earlier 7 per cent undercount of black urbanization in 1980 due to its peri-urban location, the figure of 10 per cent is not much more. What I am suggesting, in other words, is that South Africa is not really as under-urbanized as it would appear – it may well be that apartheid legislation has not so much prevented black urbanization as caused it, in large part, to occur in displaced forms.

Current urbanization policy

I am now going to attempt to describe current urbanization policy; the political–economic debate regarding its purpose is taken up in the next section.[5]

In 1981 the government divided South Africa into eight (and later nine) 'development regions'. This represented, at last, formal recognition by government that the homelands were not and could not become economically (and therefore politically) independent (see Map 2.2). These nine development regions cross homeland borders and, in certain cases, parts of a single homeland have been allocated to different regions. Clearly, criteria other than the potential independence of the homelands were used when devising these regional borders and speculation about the criteria continues. One view is that the development regions form the basis of an as yet unconfirmed, federal constitutional structure (Cobbett *et al.* 1985). This position was recently strengthened by the government's statement that it intends to create regional bodies for blacks outside the homelands which are based on the nine regions.

Map 2.2   South Africa: development regions and homelands

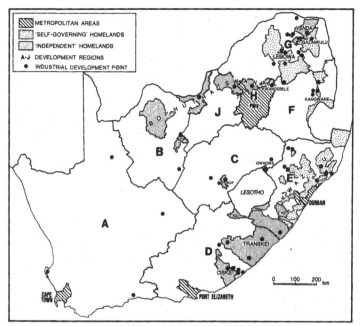

Another, complementary, view is that the development regions comprise distinct labour supply regions, in both macro- and micro-regional terms. An example of the development of a macro labour supply region is the incorporation of the northern half of the Transkei into Region E. Region E otherwise would be the province of Natal, which includes the Zulu homeland. I was informally told that opposition to incorporation from both the KwaZulu and Transkei authorities was ignored because of the importance of the labourers from the Northern Transkei to Natal's economy – labourers from the Northern Transkei are highly valued for their willingness to accept lower wages than the Zulus would accept.

The notion of a micro labour supply region warrants further analysis. There are two issues here: one is the economic structure, in particular, of the Central Witwatersrand, South Africa's largest urban centre. Between 1950 and 1970 it lost 200,000, mostly migrant, mining jobs but gained 300,000 jobs in the manufacturing and tertiary sectors during the same period. As the manufacturing sector gains little benefit from an oscillating migrant labour system aand experiences difficulties related to training and retaining a skilled labour force, urban capital has protested to this effect since the 1930s and 1940s (Lipton 1985: 150). In addition, neither the manufacturing sector nor the service sector see any advantage in a dispersed form of labour supply, which represents the loss of a market for urban household goods and services.

The second issue is the historic change in the interests of the agricultural and mining sectors that has occurred over the last two decades with respect to the migrant labour system (Lipton 1985). A simplified summary of the change is that hitherto both sectors were dependent on a supply of unskilled, cheap labour; and the low cost of this labour, before the collapse of the homeland economies, was assured by the migrant labour system. For the mining sector, the labourers left their families in the homelands or neighbouring countries, thereby forcing those economies to bear a part of the reproduction cost of that labour. For the agricultural sector, competition with urban wages was averted because blacks were not allowed to leave the rural areas if a labour shortage occurred.

Following the NP electoral victory in 1948, various factors initiated the subsequent rapid decline of the homeland economies: the removal to the homelands of the 'surplus' population from the towns; the eradication of 'black spots' (farms owned by blacks outside the homelands); and, since the 1970s, labour displaced from farms (Simkins 1981). At the same time, the war in Mozambique and the economic decline of Lesotho prevented the creation of alternative economic opportunities for foreign labour. The farms and mines enjoyed a surfeit of cheap labour. Now, with mechanization on the farms, and the mining companies introducing increasingly mechanized, skill-intensive labour processes, paying higher wages, which attract a greater balance of South African rather than foreign labour, and seeking to increase the proportion of their labour supply that is 'locally settled', migrant labour is in the process of becoming inappropriate in still more areas of the economy.

Although the trend is toward accelerating corporate opposition to apartheid in all its manifestations (Lipton 1985: 6), this is not the point. Rather, the issue is whether or not the private sector supports mobility restrictions on blacks. In this respect, all major employer organizations oppose such restrictions.

The counterpart to those pressures is the poor condition of the homeland economies. In 1983 I reviewed a development plan for the Gazankulu homeland and discovered that the decline of its economy had proceeded to such an extent that the agricultural component presumed the prior removal of 89 per cent of the people from the land to provide the balance with a poverty–line household income of R2,400 (about $1,000) per annum (Tomlinson 1984). In order to escape this need for urban migration, the government instituted an industrial decentralization programme in 1960. That the policy failed is shown by the fact that, in contrast to the more than 100,000 new entrants to the homeland labour force each year, only 34,900 jobs were attributed to the policy between 1970 and 1978 (Maasdorp 1982). Blacks clearly have had good reason to seek access to the cities.

Simkins (1981), for instance, has estimated that the annual industrial wage was three times the average annual agricultural output of a household in the homelands in the 1920s; and from the 1930s to the 1950s the difference was equal to a factor of five. However, the urbanization of the black labour force contradicted fundamental precepts of apartheid.

> As the apartheid era progressed, the social engineers looked for ways to limit the number of blacks who stayed overnight in the urban areas, either as residents with their families in township houses or as migrants in single-sex hostels. Their aim was to build up a class of 'commuters', blacks who could be allowed to work in South Africa so long as they shuttled back to their 'homeland' every night.   (Lelyveld 1986: 12)

The demise of influx control and primary reliance on the migrant labour system are therefore only a formal recognition of what is already a *fait accompli* – a distorted form of peri-urbanization. The reforms of 1986 represent no more than an additional step toward countenancing the urban shift of the black population. Although it is still considered desirable that increments to the urban black population be kept within the homelands, there is now an acceptance that black townships, located on the fringes of the city, will serve the purpose of preserving social control. Thus, the average home to work distance in India is 8 km, 11 km in Western Europe and 18 km for urban blacks in South Africa (*Bulletin Transport Information*, 1988, 2, 2: 9) – but the government intends to further emphasize the creation of a spread-out city structure. 'Future urbanisation linked to industrial development should be spread over larger geographical areas and therefore more evenly, with a reasonable degree of stimulation of urban development in the outer peripheral areas of certain metropolitan areas and large towns and not so much in remote border areas' (South Africa 1985: 5.29).

Two means of attempting to direct black urbanization to com-muter zones outside the cities are policy implementation regarding the supply and location of housing or land for housing, and the industrial decentralization policy. The focus in the latter has changed from industrial decentralization to distant locations to industrial deconcentration on the fringe of the city.

As the policy regarding the supply of land and housing is more complex, I will try to explain it first. Map 2.3 illustrates the additional land made available for blacks on the PWV in June 1988, and it shows that most of the land will be provided by filling in areas adjacent to existing townships and that stretch away from the city centre. The actual land area amounts to 13,000 hectares, but this should not be taken at face value since some of it is located in geologically unsuitable areas and/or over Johannesburg's major underground source of water, with the result that it is unsuitable for residential development. The government is, in an ongoing manner, allowing the further designation of land for black residential development, but in inappropriate locations and not yet in sufficient quantities. Thus in 1988 the government stated that the 13,000 hectares, together with another 16,000 hectares which had been allocated since 1985, would suffice until the year 2000, but were all the land developed it would provide only enough sites for about 2 million people. The probable increase in the black population of the PWV is about 5 million persons. The government's attention is focused now on ensuring that blacks locate at 'approved sites' within these developments, and an armoury of legislation, including the Slums Act and the Prevention of

Map 2.3 Land made available for black housing on the PWV

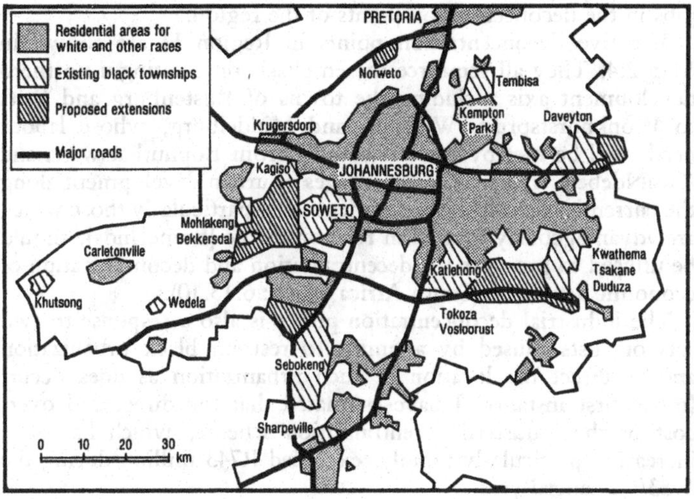

Illegal Squatting Act, has been enacted for this purpose. Because of the consequent scarcity of land, prices will rise and land will be accessible only to the relatively well off. A large proportion of the black population will still be unable to gain access to land and will be forced either to squat illegally or to rent rooms and be subject to massive overcrowding or to locate in the Bophuthatswana and KwaNdebele homelands.

Thus, housing policy replaces influx control through the imposition of controls on the availability and the location of approved sites and, with this, there is a shift of government efforts from the supply of housing to the concentration, instead, on the supply of infrastructure and the provision of finance for site-and-service schemes. Limitations on the availability of approved sites in the PWV and the ready availability of such sites in more distant settlements, accompanied by the implementation of the squatting, health and vagrancy legislation, ensure – in a less overtly racist fashion – that blacks are coerced to live at some distance from the cities.

In addition to the housing policy, the industrial deconcentration policy is intended to reinforce a deconcentrated form of urbanization by locating future industrial growth at dispersed points within about an hour's drive of the cities, in close proximity to areas of displaced urbanization. In *A Spatial Development Strategy for the PWV Complex* (South Africa 1981) it was projected that, between 1978 and 1990, 46,000 new jobs that would normally locate in the PWV would be diverted to distant industrial development points and that, between 1990 and 2000, another 110,000 jobs would be similarly diverted. It further projected the creation of an additional 300,000–350,000 jobs in the deconcentration points of the region.[6]

The five deconcentration points in Region H are shown in Map 2.4. They all reinforce the emphasis on creating a northern development axis including the towns of Rustenburg and Brits to Bronkhorstspruit, Witbank, and Middelburg, whose labour needs will be met by persons coming from Bophuthatswana and KwaNdebele. Thus, 'The advantages of urban development along the currently identified development axes, particularly those which are advantageously located in relation to the [homelands], should be utilised to promote the decentralisation and deconcentration of economic activities' (South Africa 1986b: 6.13.10).

The industrial deconcentration policy is also a response to two sets of costs caused by attempts to restrain black urbanization and to direct the location of such urbanization as does occur. In the first instance, I have estimated that the direct and overt cost of the industrial decentralization scheme, which has been increasing particularly rapidly, exceeded R743 million during the 1983/4 financial year.

Map 2.4 Deconcentration points in region H

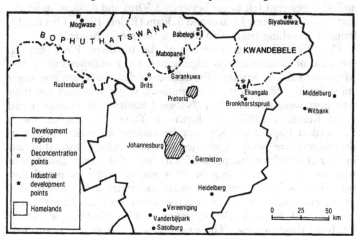

A further R1 billion was spent on having to subsidize blacks for the transport costs that resulted from inefficient residential locations (National Institute of Transport and Road Research 1985). For example, the average subsidy per black-labourer bus passenger in the Pretoria region in 1985 was R603 per year (see Table 2.3); this figure, in comparison, was equal to the per capita incomes of many individuals in the homelands in the same year. I have been informed that in 1985 the equivalent subsidy for commuters from KwaNdebele was R1,600 per annum. Moreover, the human cost of dispersed urbanization is evident from the commuting times. With respect to commuters located in the southern (and closest) part of

Table 2.3 Average annual bus passenger subsidies, 1985

City	Approximate subsidy[1] (R)
Pretoria	603
Bloemfontein	420
Durban	283
Witwatersrand	165
Port Elizabeth	72

[1] R1 is equal to about $0.40.
Source: Naude *et al.* (1987: Table 2.2).

KwaNdebele: 12.8 per cent left home between 2.30am and 3.00am and 45.9 per cent left home between 3.30am and 4.30am, with the rest leaving between 4.00am and 5.30am (Ehlers 1982). Obviously, lengthy travelling times also imply a late return home. The effect that the structure of the northern PWV has on the cost of commuting is highlighted by a comparison of bus passenger subsidy amounts for different cities. Total bus passenger subsidies in 1983/4 were R348 million, while subsidies for train transport were R866 million (National Institute of Transport and Road Research 1985). As shown in Table 2.3, the city with the second highest subsidy was Bloemfontein, where blacks in the vicinity are being channelled to Botshabelo, a township and industrial development point 55 km east of the city. Botshabelo was set up as a township in 1979 and by 1988 its population was about 450,000 persons. Future industrial development tied to Bloemfontein is designated for Bloemdustria, an industrial area located between Bloemfontein and Botshabelo, about 20 km from Bloemfontein. This example well defines deconcentrated urbanization: the creation of settlements, deliberately dispersed for the purpose of social control but which, despite their isolation from the cities, are intended to provide urban labour.

In contrast to Bloemfontein or Pretoria, Port Elizabeth has a much more concentrated urban structure and a subsidy of only R72 per passenger. However, it was at Langa, outside Uitenhage near Port Elizabeth, where the 1984 massacre of blacks proceeding to a funeral sparked a higher intensity of revolutionary struggle. At that incident, the police were reported to have become flustered because of Langa's proximity to white residential areas and the possibility that the blacks could enter those areas (Haysom 1985). The type of spatially extensive urban form engendered by the country's urban policies is intended to counteract such possibilities.

Clearly, the apartheid city is an extraordinarily inefficient and expensive city, both for the government and for the majority of its inhabitants, whose quality of life, productivity and material circumstances are negatively affected. Recent limited willingness on the part of the government to increase the supply of land on the urban fringe denotes a move away from this situation, but the question is: how far will the government go in addressing the land needs of the really poor who are unable to afford housing?

The purpose of the urban and regional policies

Papers forming part of a workshop (whose proceedings were edited by Tomlinson and Addleson 1987) maintain that the government's urbanization policies, in addition to being intended to improve the government's legitimacy and control, are also intended to serve the interest of (an undifferentiated and abstract) capital (Cobbett *et al.* 1987; Hindson 1987b). While I agree with the first position, I cannot see why the policies serve capital or, more exactingly, why they serve capital better than the obvious alternative policy of unrestricted urbanization. Examples below of housing policy and the location of black settlements will be used to illustrate this difference.

Hindson (1987b) explains the housing policy as follows: the government's persistence with restrictions on the location and supply of land and housing is intended to maintain state authority over communities resisting influx control as well as to ensure residential segregation and the settlement of blacks at some distance from the cities. Although influx control has been done away with and residential segregation is leaky, the general thrust of his argument is unobjectionable, namely that the policy is directed at safeguarding the government's efforts at social control. The issue is not, as is commonly thought, one of an Afrikaner racial aberration, but rather the need the government feels to protect centres of production and areas of white residence. Historically, Group Areas legislation served this purpose. Now, with the demise of influx control and in anticipation of the removal of racial zoning, a deconcentrated urban form must suffice. When considering government reforms therefore, it is important to bear in mind that the urban form of the apartheid city is complete in most cases[7] – there have been 3.5 million forced removals (Surplus People's Project 1983), neighbourhoods have been torn down and the cities have been racially remodelled. It follows that when the NP states its intention to undertake incremental reforms, this does not mean that integration of the cities will result. If residential location was left to the market, the majority of blacks would be forced by increasing land prices (which result from the limited zoning of residential land and supply of infrastructure) to locate outside the city.

A location outside the city also makes possible very specific forms of social control. For example, McCarthy and Swilling[8] point to the difference in the nature of political action emerging from squatter settlements and formal townships. In Natal, opposition to the government is most pronounced in the formal townships, where the government owns the houses and supplies the services; the same is not true in the squatter settlements, where land and the services that exist are supplied through 'warlords' belonging to Buthelezi's 'party', Inkatha. Inkatha, of course, has for some years been engaged in a deadly battle with Cosatu and the UDF.

Similarly, the government enlisted the support of vigilante forces at Crossroads (a squatter settlement outside Cape Town) to defeat, again with considerable loss of life, the Comrades, who are allied to progressive movements. Hindson (1987b), however, takes the analysis of current housing policy further, with specific attention to the acceptance of site-and-service housing schemes. His view is that these schemes are intended to lower the cost of reproducing the labour force in order to reduce the upward pressure on wages. In itself, the point seems unobjectionable but, when measured against a more exacting question suggested by Fainstein and Fainstein (1985), there is considerably less clarity. If reducing the reproduction cost of labour was, in fact, a goal of government policy, it could be achieved more effectively and economically simply by eliminating restrictions on the supply and location of land for housing and, similarly, by eliminating building restrictions. That the government selects an option that is cheaper than formal housing, is largely a response to the fiscal crisis. Being unable to provide housing, as it did in the past, it now takes a less costly, but not the cheapest, option: although site-and-service schemes ensure a cheaper supply of housing, in projects undertaken around the world subsidies have often exceeded 50 per cent of the cost of the project (Rodwin and Sanyal 1987). Why was a barrier of legislation enacted to prevent illegal squatting and to safeguard health standards? Why didn't the government simply facilitate self-help housing processes? Hindson's error is that he has confused a consequence of the new housing policy with its initial motivation.

The argument of Cobbett *et al.* (1985, 1987) and Cobbett (1987), that the creation of labour supply regionns is a benefit to capital, is also difficult to substantiate. For example, the urban policies represent the government's 'reconceptualising spatial forms, in order to facilitate and manage the development of new patterns of domination, exploitation and social reproduction', but why do these policies 'provide a spatial framework for the renewal of capital accumulation' (Cobbett *et al.* 1985: 88, 89)? Regarding the reproduction cost of labour, why should labour, instead of being coerced to live in a commuter zone some distance from the cities, not be allowed to reside where work is available? Surely, unrestricted urbanization would reduce the cost of housing and transport, and minimize the commuting time of labour.

Given that the greatest material benefit for capital would result from unrestricted urbanization, it is a mistake to persist with the view that something less than this is in the interest of capital. It should also be noted that restricted urbanization has 'political' costs because it enables the South African left wing to continue to argue that the relationship between capitalism and apartheid is collaborative (Saul 1986). Among many unions, the ideological battle has been lost (or won, depending on one's persuasion) (Legassick 1985) – capitalism is associated with apartheid and the resultant view is that to abolish apartheid one has to struggle for socialism. It would serve capital well if individual liberty could be achieved independently of the socialist struggle, thereby severing this perception of the relationship.

The disjuncture between urban policy and the interests of capital can be explained thus. There are two reasons why the government formulates policies that serve capital. The first is universal – governments depend on income derived from the taxation of profits and wages (Offe 1978). The second reason is specifically South African – it is only since the late 1960s that an Afrikaner business class, ardently nurtured by the NP since 1948, has emerged (Lipton 1985). A bifurcation in state interests now occurs because, although most Afrikaners earn their livelihood directly through working for the state or state-owned enterprises, there is at the same time a powerful class of Afrikaner capitalists (Charney 1984). The latter, however, are not altogether independent of government since 'much of Afrikaans private business, often the weak sister of its English competitors, remains dependent upon state patronage and protection, while the agricultural bourgeoisie relies on farm pricing, input, and credit policies' (Charney 1984: 281).

A year or two ago one might therefore have confidently asserted that Afrikaners fail to distinguish between their class and ethnic interests – the two are fused (Slabbert and Welsh 1979); that Afrikaners have good reason to insist on white, *de facto*, Afrikaner control of the state; and that in this crucial respect they are distinct from English liberals and the predominant owners of capital, whose material self-interest is relatively independent of the government. Now, the extent to which a class of independent Afrikaner bourgeoisie has arisen, itself reflected in the decline of Afrikaner solidarity, helps explain the government's reforms. Thus, in general, government policy can be expected to favour capital but, in specific policy instances, the government, still unable to shed its apartheid and ethnic interests, may choose to pursue its own, different, concerns. My view is that there is a tension between ensuring conditions favourable to capital accumulation and pursuing ethnic policies, that this is especially manifest in urbanization policy, and that the slow whittling away of racist urban policy will follow the redefinition within the NP of what constitutes 'tolerable' integration.

To sum up, the situation in South Africa is that the state is constrained by the prerequisites for capital accumulation but, at the same time, it may consider in a calculated manner that it is necessary in certain instances to act against those prerequisites. Urbanization policy presently represents just such an instance.

Conclusion

The extent of black urbanization has proved difficult to pin down – a figure in the late 1980s of about 50 per cent is credible. Apartheid practices have restrained black urbanization but such distortions are likely to be less common in the future. In their place, however, is the attempt to create a deconcentrated form of urbanization, with black residential proximity to the city being conditional on the ability to afford urban land and services. As long as residential segregation lasts, this will consist of spatially distinct black townships located on the outskirts of the cities. Once statutory residential segregation is removed, the urban form will remain, but now directed market forces will serve to restrict the access of the poor, *de facto* black, population to the cities.

Until the post-apartheid era arrives, the urban poor will be ill served by the current urbanization strategy. Three examples will demonstrate the point. First, the new mechanism of influx control has become more than simple controls on the supply of land and the resultant high prices. Where land and housing are availablee, the effective restraint to closer urbanization will be overcrowding. The poor will be confronted with the choice of overcrowding or living further out and spending a larger proportion of their income on commuting rather than on food, education or shelter. Overcrowding also serves to exacerbate psychological distress and diseases associated with the lifestyle. Second, deconcentrated urbanization entails enormous commuting distances – for tens of thousands of blacks the costs to comfort, family life and productivity are extreme. Finally, a deconcentrated form of urbanization inhibits the possibilities the poor have for making a living. For example, when low levels of demand are concentrated in a large urban centre, the size of the market is increased and informal sector opportunities are enhanced; but these same opportunities are reduced when that demand is spread out among the townships. Generally speaking, because the townships represent the isolation of the poor from the relatively wealthy, there is a lack of economic opportunities and a lack of services and recreational facilities.

But none of this is inevitable or necessary – there is considerable opportunity for concentrated development in South Africa's cities and little scarcity of land. For a post-apartheid government the issue will be how best to ensure that land is made available for residential purposes and how to facilitate the housing supply; also how to do both in such a way that the linkages between work, residence and commerce are convenient and efficient. Land and housing are not issues that can be distanced from metropolitan structure.

Notes

1 Some persons who have been involved in union or other opposition activities have had their permit to live in the rest of South Africa withdrawn, and they have been forced to go to 'their' independent homeland.

2 Deconcentration means the creation of a spatially extensive urban form. This is achieved by the location of residential development at unnecessary distances from the city centre and work places, and through industrial location policies intended to attract investment to sites about 50 km outside the city.

3 The term 'federal' embraces the notion of 'own' affairs being decided at the regional level, and 'general' affairs being decided on at the central level.

4 Simkins later equivocates and concludes with an estimate that 1.5–3 million persons are being kept out of the cities.

5 Much of the material for this and the next section comes from Tomlinson (1988).

6 Although the *Spatial Development Strategy* refers to deconcentration points, the points mentioned in the Strategy are not entirely the same as those indicated in Map 2.4. The points specified in the Strategy are Springs-Brakpan, Bronkhorstspruit, Rosslyn and Brits.

7 This was a point made by W. Cobbett during a lecture given in the Department of Town and Regional Planning, University of the Witwatersrand.

8 In conversation.

3 Government and economy in a democratic South Africa[1]

My reason for hypothesizing a post-apartheid state is to set up a context within which urbanization policies can be considered – the nature of the state will obviously determine the relevance of alternative policies. There is, though, another side to this coin: all governments serve distinct constituencies and, since this book addresses the interests of the urban poor, it is illuminating to debate whether a government led by the African National Congress would adopt policies beneficial to the poor. In this regard, to call an ANC-led government 'progressive' is to stop short of noting that such a government is likely to have two primary constituencies – petty bourgeois groups and organized labour – neither of which will form part of the poor. Thus, when the Freedom Charter urges that the 'Slums shall be demolished, and new suburbs built. . .', this is distinctly a policy whose expense will ensure that it excludes the urban poor. In other words, the novelty of a democratic government should not blind one to the fact that even democratic governments promote selective redistribution.

The government hypothesized is an amalgam of what is, at present, a coalescing of views within two competing images of the post-apartheid state. I have presented the images in detail because, as I stated earlier, my presumption is that this topic has intrinsic interest for the reader.

The chapter has been divided into four sections. The first portrays the federal and capitalist image, and the second presents views within the ANC and its allies. The Freedom Charter, which the ANC and many others honour, is examined and found to be a document that is admirable in sentiment but lacking in specificity. Consequently, the ANC's more recent Constitutional Guidelines are considered alongside the Charter. The Guidelines are helpful, but apparently represent a draft intended to foster debate for their further development. A complete constitution is not being proposed at this stage by the ANC – this awaits democracy and the participation of the people. Greater clarity regarding the likely nature of the post-apartheid state is achieved through looking at the socialist debate in the unions and the differences between African nationalists and the South African Communist Party. The third section examines the likely relations between capital and the ANC, the former's motivations for a federal system and the latter's desire to minimize constitutional impediments to the expression of majority views. It is in the conclusion that I hypothesize the likely nature of the post-apartheid state and offer a few supporting arguments.

Capitalism and Federalism

There are a number of portrayals of the federal option for South Africa. In addition to Lijphart (1985), there are texts by Beckett (1986), Boulle (1984), Louw and Kendall (1986) and Slabbert and Welsh (1979). The decision to refer to the National Party and to Lombard and Du Pisanie (1985), and to the Progressive Federal Party and the Democratic Party, reflects the fact that I am more concerned with the views of influential South African actors than with those of academic commentators. (Since the PFP has joined the DP, I refer to the DP where possible. However, the DP's views are still inchoate and the PFP, whose former members predominate numerically within the DP, still provides a good measure of the liberal cause.) The additional reference to Lijphart (an observer of South Africa and author of a number of treatises on constitutional matters) reflects both his influential role in articulating the options in South African and his outlining constitutional principles that form a useful basis for the subsequent analysis. Lijphart, in fact, advocates consociationalism, and federalism is viewed as a type of consociationalism. An examination of the principles of consociationalism is especially useful since it highlights the options underlying the federal debate.

A common characteristic of many of the depictions of federalism is that the need for associated economic changes is disregarded. Despite examples such as Yugoslavia and the Soviet Union, whose constitutions demonstrate that federalism is not necessarily associated with capitalism, the search has primarily been for political solutions. Indeed, some portrayals actually go beyond simply ignoring economic options and actively minimize the relevance of choice with respect to the country's economic arrangements. Beckett (1986: 67), for example, argues with some abandon that 'The problem is one of a political system and not an economic system. No economic system can work unless the political foundation is sound. Whereas conversely if the political foundation is sound then it doesn't matter what the economic system may be.'

Another example is Louw and Kendall (1986), whose work is notable for having had a foreword by Winnie Mandela. They are entirely disingenuous when they argue that public sector intervention in South Africa's economy is too pronounced – one needs a free market economy – and then persist with the view that the choice between capitalism and socialism is, anyway, not an issue. Having posited a prior context of a central government with extraordinarily limited powers and a Bill of Rights that enshrines property rights, Louw and Kendall would have one believe that the socialist option has not somehow been impeded. The option, they argue, remains available as a result of their constitutional design. A limited central government, accompanied by extensive decentralization based on the Swiss canton system, supposedly enables each canton to select the economic system it desires. An ANC-dominated canton, they suggest, can opt to be socialist and a PFP-dominated canton can choose capitalism!

A last note, prior to considering Lijphart, concerns the suspicions of many that federalism entails an underlying agenda. Supporters argue that a federal system of power-sharing is the system of government most likely to be accepted by the opposing forces in South Africa. They further maintain that in deeply divided societies federal systems are more likely than unitary systems to restrain competition over resources to means specified in the constitution. This leads to the additional point that civil rights are thought most likely to be maintained within a federal system. For those concerned about the ANC's democratic intent – for example, when pressed at the Mfuwe Lodge discussions[2] regarding which industries the ANC would nationalize, Thabo Mbeki remarked that the press would be the first – federalism offers the additional benefit of constitutional safeguards against the capricious exercise of power in the manner of the NP. Constitutional safeguards, however, may also be viewed as an attempt to impose constraints on the expression of majority views. In particular, a federal government founded on capitalism would subsequently find it extremely difficult to adopt a socialist incline. There may be good reasons for federalism, but, equally, there are good reasons to be suspicious of it.

Lijphart

In Lijphart's view the PFP's constitutional proposals are fully consociational, and, indeed, even the NP has had occasion to refer to its plans as consociational. The principles on which consociationalism is founded are at the centre of the political debate among whites. While one might say that the debate is therefore somewhat academic, no discussion of white reform ambitions should proceed without reference to them. I thus begin with the principles, as Lijphart (1985: 6) describes them.

The four basic elements of consociational democracy are (1) Executive power-sharing among the representatives of all significant groups; (2) A high degree of internal autonomy for groups that wish to have it; (3) Proportional representation and proportional allocation of civil service positions and public funds; (4) A minority veto on the most vital issues.

Lijphart (1985: 5) juxtaposes consociationalism with the Westminster system because a 'Majoritarian democracy – that is, a democratic political system without special autonomy and protection for ethnic and other minorities and without guaranteed minority participation in governmental decision-making – is both unfair and unworkable in a society as deeply divided as South Africa's.' Consociationalism is preferred since majoritarianism means, in effect, that the will of the majority prevails even where, in practice, minorities have not had a chance to participate in the decision-making. Furthermore, in heterogeneous countries, where voting behaviour is based on adherence (linguistic, ethnic, religious) to a particular segment, the minorities are permanently excluded from a role in decision-making.

Power-sharing is the most important of the four basic elements of a consociational democracy. It refers to 'government by a broadly inclusive coalition' (Lijphart 1985: 6); and can 'take various institutional forms, such as that of a grand coalition cabinet in parliamentary systems, a grand coalition of a president and other top office holders in presidential systems, and broadly inclusive councils or committees with important advisory and coordinating functions' (1985: 7). This explanation is, no doubt, sufficiently vague to create immediate misgivings in many readers but, before discarding consociationalism, the reader is referred to the more specific PFP proposals, which indicate how such a constitution might work.

Segmental autonomy is consociationalism's second element. Following Adam (1983), the social cleavages Lijphart anticipates are comprised of Afrikaners and many coloureds, blacks with some coloureds and a few whites, and English-speaking whites with most Asians and some coloureds and blacks. The black segment is thought likely to split along Zulu–Xhosa lines. Lijphart's (1985: 7) view, one that has disquieting similarities to the NP's notion of 'own and general affairs', is that 'on all issues of common interest, the decisions should be made jointly; on all other issues, each of the segments should be allowed to decide for itself'. Where segments are geographically concentrated, territorial federalism is appropriate. Where the geographical concentration is absent, corporate federalism, based on segments identified by voting patterns, is suggested. The latter point is critical. When referring to segmental autonomy, Lijphart does not presume any prior predetermination of the segments. South Africans, hitherto accustomed to and burdened by the government's racial classifications, would through voting patterns, define their own segments.

Proportional representation is the third essential element; 'both majorities and minorities can be "winners" in the sense that each group is able to elect candidates in proportion to its electoral support' (Lijphart 1985: 8). However, proportional representation need not mean that everybody's vote is equal. For example, in the American lower house, each congressperson represents an equal number of constituents, but, in the upper house, each state has two senators regardless of the number of people in the state. The result is that the vote of one Alaskan is worth the vote of 100 Californians when senators are elected. In this manner, both protection and participation are afforded (territorial) minorities during the legislative process.

Finally, more complete protection is afforded minorities by the veto. This means that a party that gains x per cent of the vote is able to exercise a veto, either absolutely or in a suspensive sense. Thus, in the Namibian elections for representatives to participate in the drafting of the country's constitution, the South West African People's Party had to win a two-thirds majority in order to have a free hand in writing the constitution – one-third of the vote represents a veto. This veto power obviously also acts as an incentive for power-sharing and the formation of coalition cabinets.

The primary quality of consociationalism, Lijphart argues, is not that it is a superior constitution per se, but that in South Africa it is the system most likely to be acceptable to the opposing sides in the struggle for power.

The Progressive Federal Party and the Democratic Party

The DP, and the PFP before it, support federalism. The arguments for this position have a coherent base – one articulated by Slabbert and Welsh (1979) – and I will present their views before proceeding to an examination of the PFP's and the DP's constitutional proposals.

Slabbert and Welsh (1979: 24) note:

For as long as white domination persists ethnicity will be a less salient feature in the political arena. In fact there will be a strategic black solidarity against white domination. However, we believe it would be very shortsighted to ignore the possibility of ethnic competition and conflict once white domination has been ended.

Ethnic conflict is likely when there are distinct inequalities in society and when recourse to ethnicity enhances the ability of certain groups to gain superior access to the allocation of resources. In this light, English-speaking white South Africans, who are a minority ethnic group, would be expected to disavow ethnicity, whereas majority ethnic groups, like the Afrikaner with the NP, or the Zulu with Inkatha, are likely to emphasize it. In other words, Slabbert and Welsh do not base their views on crude notions of tribalism. Instead they stress that ethnicity is a 'relational concept', one that emerges when it becomes a convenient way for groups to seek advantage.

This view represents an obvious point of debate with the Marxist anticipation of the future primacy of class rather than ethnic conflict. They consider the Marxist view at some length and argue that for most blacks the issue of race and class is fused. In this respect Nelson (1979: 219) is especially perceptive when she observes that,

> since ethnic identities have persisted, the complicated and highly variable pattern of emerging class and occupational identities in plural societies is superimposed on and interacts with ethnic identities. The expectation that class ties will replace ethnic links is therefore doubly simplistic: in its implied definition of both class and ethnicity as single, simple, and static social identities, and in its assumption that the two cannot coexist. The relevant question, then, is how emerging, often vague, and intermittent class identity interacts with, rather than replaces, persistent ethnic identities.

Thus, while black emphasis on the rejection of ethnicity as a means of opposing apartheid is laudable, it should not lead to any expectations regarding the coherence of black political action after apartheid's demise. Slabbert and Welsh (1979: 95–6), therefore, continue with the view that

> No assumptions can be made about the potential distribution of black voter allegiance in a future open election in which universal suffrage prevails and in which presently banned black

organisations are able freely to campaign and mobilise support. Almost certainly sharp policy differences would crystallise among the several political parties that would be bound to arise. Black nationalism in South Africa has never been a monolithic entity; rather it has been a congeries of groups with different ideological, class and perhaps, regional interests, held together by a common (predominantly defensive) rejection of racial discrimination.

Like Lijphart (1985), Adam and Moodley (1986) and, indeed, many other commentators, Slabbert and Welsh (1979) agree that the possibility of conflict in divided societies is exacerbated by a unitary government. If the divisions they anticipate, or the cleavages predicted by Adam (1983), come about, they hold that a federal constitution will serve South Africa best. This is not to say that ethnic or other conflict will be averted in a federal system, for ethnic conflict is found in any constitutional arrangement. Rather, it is to say that the above authors agree that a federal system is likely to avert more conflict than any alternative system. Federalism facilitates the devolution of responsibility for politically abrasive 'own affairs', such as schooling, and redirects conflict so it is not for control over the state per se, but rather for control over regional and local government. In addition, federalism, with an American-style upper house, ensures over-representation for minority views, which brings them into the central decision-making process.

Happily, the PFP did not, and the DP does not, prescribe South Africa's future constitutional structure. They recognize that blacks can no longer have constitutional systems foisted upon them and, instead, call for a national convention wherein a constitution would be negotiated. The DP's constitutional proposals, at the time of writing, have advanced little beyond the detail necessary to fill an election manifesto and, because they are apparently largely based on those of the PFP, I will proceed through description of the PFP's proposals and occasional cross-reference to the DP.

The PFP proposes 'a federal and decentralised system of government, a bicameral federal parliament with a proportionately-elected lower house and a senate representing the states, a power-sharing federal cabinet, and a strong minority veto' (Lijphart 1985: 67). The lower house will elect the prime minister by majority vote, so the prime minister will either represent the majority segment or a coalition of segments. The prime minister's selection of cabinet members will reflect the strength of the various parties in the lower house and will obviously involve negotiation with the leaders of those parties. All parties with at least 10–15 per cent of the vote will be entitled to representation in the cabinet.

The continuity from the PFP to the DP is evident in the latter's call for 'universal adult franchise in a federal system'.

> Each party would be represented at all levels of government in proportion to its support. If a party gets 10% of the vote it gets 10% of the M.P.'s. [Members of Parliament] Significant parties will be represented in the Cabinet where they will be obliged to negotiate with each other before laws are passed, ensuring a government that takes the opinions of all South Africans into account. (DP 1989)

The PFP's proposal for segmental autonomy is striking because it involves both territorial and corporate federalism. On the one hand, 'There will be a massive devolution of power, along federal lines, to regional and local governments' (PFP 1986). On the other, each party or segment may establish a cultural council, with a view to pursuing its interests, and will have the power to run schools and cultural associations, much as private schools operate in South Africa or the Catholic Church operates schools in the United States. Each cultural council is further empowered with the right to elect a senator.

There are various more detailed features of the proposals, such as three types of minority veto, but these are not especially significant here. The important point is that persons such as Archbishop Tutu have spoken favourably of the PFP's proposals (cited in Lijphart, 1985: 73). While I do not anticipate that Tutu would ever oppose the ANC or the UDF, his interest suggests that such federal proposals represent a realistic starting point for negotiation for liberal and capitalist interests.

None the less, there are at least two queries worth making about these proposals. The first concerns the right of cultural councils to elect senators. This principle would seem to intensify ethnicity since any self-respecting ethnic entrepreneur would be expected to attempt to form a cultural council and to advocate ethnic politics. It may be that this line represents a sop to Afrikaners, who would not form a majority in any likely definition of the federal regions. But I struggle to see the value of a constitution that makes provision for, and so exacerbates, ethnic divisions, and instead see considerable reason for pursuing an individual Bill of Rights through which people can come together voluntarily to develop religious, linguistic, schooling or other interests.

The second query concerns the continuing vagueness of the PFP's, and now the DP's, economic policies. Slabbert and Welsh

(1979) have argued that material inequalities are central to understanding ethnic and other divisions, but the PFP's proposals address
the form, rather than the causes, of the resultant competition. The
DP is clearer, albeit in intent more than in detail, when it advocates deregulation, free enterprise, attracting foreign investment,
lower taxes and streamlined civil service. However, at the same
time, the DP urges greater expenditure on health, education,
old-age pensions and housing. This begins to look like a manifesto
drafted by Reagan and one wonders whether the DP and its
business allies have begun to confront Nolutshungu's (1982: 16)
scepticism.

What is clearly lacking from the liberal debate on reform is
any systematic concern with the question whether the capitalist economy can, even in economic terms alone, meet the
demands of reform or its consequences . . . What is absent
is a recognition along the lines of James O'Connor's *The
Fiscal Crisis of the State*, for example, which sees reform as
an economic necessity realised through the political process
of liberal democracy but one which brings in its train a
series of economic difficulties which the economic base cannot
resolve.

Le Roux (1986) indicates that debate on this score has, in fact,
now arisen and reports that some economists are not entirely
unoptimistic, provided that there is rapid economic growth! Failing
these conditions, it may well be that capital's interests would be
better served by a unitary government, the election of a party
whose leaders have been co-opted and a drift into authoritarianism.

The National Party and Lombard and Du Pisanie

Like the DP, the NP's 1989 electoral manifesto (*Plan of Action*) calls
for federalism, a free-enterprise economy and minority veto power.
However, unlike the DP, the NP seeks to impose preconditions
for constitutional negotiations, which include group rights, and
hopes to involve homeland and other non-representative 'leaders'. In addition, the context for elections imposed by the NP
consists of political trials, political prisoners and a State of Emer-
gency. Further, the NP's intentions are coloured by an underlying
vagueness. Thus 'De Klerk has declared himself in favour of four
separate "constituent assemblies" for "own affairs" and some sort
of consensus-based multi-racial executive controlling general affairs
"without any one group dominating any other group"' (Phillips
1989: 12). For these reasons, the NP's deliberations lag so far

behind the views of the majority of the population that, if one can say it of those who control the guns, they seem trivial.

The NP lacks credibility, but there are suggestions of advances and flexibility in its thinking. For example, in its electoral manifesto, advertised in the media as a *Plan of Action*, the NP notes that 'The present basis in terms of which groups are defined for the purpose of political participation creates many problems', and continues: 'Freedom of association and of disassociation must, as far as possible, be points of departure' and 'A person must be able to change to another group'. Perhaps, at last, South Africans can look forward to freedom from racially prescribed group membership.

In many respects the first, and still the most comprehensive, depiction of NP re-thinking is the Lombard and Du Pisanie (1985) document. This document is important because it reveals the attributes of federalism and free enterprise in NP deliberations and, to this day, is an advance on most discussions on reform within the NP because of its acceptance that the Afrikaners would lose control of a considerably weakened central state. I have used their report because it comes close to what, for the NP, would constitute a serious negotiating stance, and because it therefore enables an exploration of what, from the NP's point of view, it would mean were the party to relinquish power.

It is worthwhile to digress here to further establish the credentials of their report. Lombard was formerly the prime minister's economic adviser and when the report was prepared both he and Du Pisanie were economists at the University of Pretoria – a centre stage of Afrikaner deliberations. (They have both since left – Lombard to become the deputy-governor of the South African Reserve Bank and Du Pisanie to join a commercial bank.) The Associated Chambers of Commerce (Assocom) commissioned their document with a view to formulating proposals for the *Removal of Discrimination Against Blacks . . .* for South Africa's Minister of Constitutional Development and Planning. As is now apparent, Du Pisanie later indicated that their report was being favourably considered at the cabinet level (cited in McCarthy 1986).

Assocom's response reinforces the impression that federal designs are favoured more because they help to safeguard capitalism than because they facilitate the realization of democracy. The organization notes that it '*endorses the view that economic freedom and the private enterprise ethic – as well as the norms with which they are associated – are best entrenched in a future political system embodying principles of federalism or confederalism*' (Lombard and Du Pisanie 1985: preface). The emphasis here should mostly be on 'entrenched' since federalism is intended to weaken the central government and to make it all the

more difficult to adopt policies that change the prevailing economic
arrangements and redistribute wealth.

Before describing the report, it is informative to query what role
Lombard and Du Pisanie thought their report might play. Initially,
seemingly in good faith, they note that they are preparing 'an
agenda for negotiation'. On the last page of the report, however,
they state that

> it would obviously be unwise to dismantle the existing power
> structure in one fell swoop without any idea of the nature of
> the power structure that would take its place, or the way in
> which the battle for power would be waged. . . . Evolution-
> ary change from the existing political dispensation to a fully
> legitimate new dispensation therefore requires that *the existing
> power structure should only be gradually dismantled as power can
> fairly safely devolve upon new structures* supported by the people
> involved.
>
> (Lombard and Du Pisanie 1985: 93, 94; emphasis added)

The ANC, however, is unwilling to accept a gradual move
away from NP rule, most obviously because it appears that the
constitutional structure towards which change would be directed
would be somewhat less than freely negotiated. Similarly, the PFP
recognizes, as do Lijphart and most other observers, that a freely
negotiated constitution is central to the legitimacy of any future
state. Lombard and Du Pisanie, on the other hand, evidence the
typical reluctance of the NP to relinquish 'ultimate control'. There
is a tension between their desire to direct events and their noting
that 'the primary prerequisite for political stability in a democracy
remains substantial consensus among the people when creating
the state about the rules of the game' (Lombard and Du Pisanie
1985: 3). In effect, Lombard and Du Pisanie do not advocate an
agenda for negotiation in the manner that the PFP does – they
are attempting to sell an agenda to the authorities for their largely
unilateral constitutional manipulations.

The purposes of the Lombard and Du Pisanie proposals are
found in the attributes of federalism. The first, evident from
the Assocom endorsement, is that 'The chances that the values
of individual freedom, property and contractual rights may be
upheld by black citizens of the future Republic of South Africa
will be considerably enhanced *if this Republic could rest on federal constitutional foundations*' (Lombard and Du Pisanie 1985: 25; emphasis
in original). Federalism is intended to diminish the power of the
central state relative to the secondary and tertiary levels. Together
with constitutional guarantees and an independent judiciary, the

hoped-for outcome is to be one where 'hostile political majorities' can institute neither socialism nor even 'excessive' taxation.
In the latter regard, Lombard and Du Pisanie (1985: 7) go so
far as to attempt to institutionalize additional 'basic elements of
South African civilisation . . . namely . . . the basic rules about
the standard of the value of the national currency, and . . . the
principles of taxation'. Clearly, a primary concern here is the
defence of property, but it is ironic that Nationalists, who come
from a party that, through artifice and coercion, has demonstrated
the frailty of constitutions (Yudelman 1987), should now lay such
a heavy emphasis on the need for constitutions and entrenched
clauses.

The second purpose concerns the protection of Afrikaner interests. Most Afrikaners believe that their political control is essential
if they are to retain their identity (Adam and Moodley 1986), but,
in fact, the motivation for retaining group identity goes deeper than
this. Friedman (1988: 26, 27) explains that government officials
believe that 'a nonracial democracy would ensure the election
of a black government which would both outlaw its opponents
and appropriate the economy'. Thus the government 'is willing
to accommodate black demands but only if whites lose nothing
in the process; this can be assured only if whites continue to
wield power as a group . . . This illustrates the intractability of
the problem, for the demand is not that whites retain cultural
and language rights, but that they retain political privilege based
on race.'

Thus NP policy, at least until 1989, had it that South Africa
consists of thirteen distinct nations, each of which is a minority
group. The nations are the various black ethnic groups (the Zulu,
Xhosa, Tswana . . . treated as separate nations); the Asians (Chi-
nese and Indians, Moslem, Hindu, Christian and Buddhist, being
treated as one nation); the coloureds; and the whites (Afrikaners,
English, Portuguese, Italians . . . being treated as one nation). The
NP would prefer to establish ethnic constitutional protections,
but Lombard and Du Pisanie (1985) appear to accept that such
clauses would be rejected. They nevertheless press forward with the
shibboleth of there being thirteen nations and argue for federalism
based on states defined in terms of their ethnic make-up. The
problem, in their eyes, is that whites would not form a majority
in any region. Their response is that

> Should territorially based federal states be established in South
> Africa, it is clear that *local Authorities with extensive powers cov-*
> *ering as many culturally sensitive government functions as possible,*
> will be most important in safeguarding the autonomy and

self-determination of the various population groups. The federal principles governing the relationship between the national
government and the state governments will have to be made
applicable to the relationships between all state governments
and their local authorities. Should this rule not apply, no
guarantee would exist that local communities would be able
to maintain their autonomy.

(Lombard and Du Pisanie 1985: 91)

In their original report, they fudge the issue of whether local
authorities should be racially distinct, in other words, whether the
Group Areas legislation should remain. They seem to suggest it
but, apparently, have since accepted that such legislation should
go. It is no doubt clear to them that it is largely unnecessary:
whites can afford more expensive houses and neighbourhoods
and, while some integration might occur, they would still form
majorities in many neighbourhoods. In addition, with the natural
segregation of English- and Afrikaans-speaking whites that already
occurs, Afrikaners would retain majorities in many local authorities
and would be able, for example, to retain Afrikaans as a medium
of instruction in 'their' schools.

The problem with the report is that it never achieves any legitimacy. It is difficult to perceive it as having been drafted 'for the
good of South Africa', rather than for more narrow interests, and
it has, perforce, to be understood in the light that federalism is
not viewed as good, per se. Instead it reflects a manoeuvre in the
sense that 'The ruling elite, morally stigmatized and ideologically
confused about appropriate strategies and associated values, has
become obsessed with devising ways to outmanoeuvre its adver-
saries' (Adam and Moodley 1986: 165). The agenda underlying
the negotiation process intended to safeguard white property and
Afrikaner interests ensures that the NP's, and Lombard and Du
Pisanie's, views are unacceptable to most South Africans. Lombard
and Du Pisanie, in fact, do a disservice to the federalist cause – while
Slabbert and Welsh (1979) have presented arguments for the social
value of federalism, Lombard and Du Pisanie draw attention to its
potential role in blocking majority views.

The ANC and the desire for a welfare state

There is increasing debate about the ANC's view of the optimal post-apartheid economic and constitutional arrangements. The
Freedom Charter is usually employed for doctrinal clarity, but this
serves only to shift the emphasis of the debate to the meaning of

the Charter. One must accept that much of the debate is fatuous. Leaders within the ANC are not approaching majority rule with, to use a fearful analogy, Pol Pot's theoretical convictions. During the 1985 discussions at Mfuwe Lodge, Tambo, for example, said that the details of the intended nationalization of monopoly capital in South Africa had not been worked out, and further, that the decision would be made by the people and that the SACP's view would be one among many contending views. What this suggests is that as much attention should be devoted to likely struggles within the ANC as to interpreting the Charter and the ANC's recent Constitutional Guidelines, both of which are reproduced in the Appendix. The Charter none the less remains the central ANC policy document and the following discussion of the ANC's position begins with an examination of the document, together with the later Constitutional Guidelines. The Guidelines are at this stage a bit confusing since it is unclear whether they are intended to be 'a coherent blueprint for a future society, or a tactical intervention to broaden the ANC's base, widen its appeal and accelerate progress towards a negotiated settlement' (Glaser 1988: 28).

The relationship between the ANC and the SACP and the 'two-stage' theory, through which the SACP justifies collusion with African nationalists, are considered next. While the relationship offers both sides distinct strategic benefits, it is likely to become turbulent as the 'national democratic revolution' progresses.

The Freedom Charter and the Constitutional Guidelines

Identification with the Freedom Charter is a defining characteristic of the ANC, the United Democratic Front, the Mass Democratic Movement and the Congress of South African Trade Unions, although, in Cosatu's case, the formal adherence patches over contending forces in favour of 'socialism now' or the Charter and a national democratic revolution (Nidrie 1988a).

The examination of this document provides one with an initial appreciation of ANC positions. With respect to its political content, it emphatically calls for a democratic South Africa.

THE PEOPLE SHALL GOVERN!

Every man and woman shall have the right to vote for and to stand as a candidate for all bodies which make laws; All people shall be entitled to take part in the administration of the country; The rights of the people shall be the same, regardless of sex, race and colour; All bodies of minority rule shall be replaced by democratic organs of self-government.

It is less clear, however, that the Charter calls for a unitary state. For example, the Charter is first phrased not in terms of individual rights but, rather, in terms of group rights!

ALL NATIONAL GROUPS SHALL HAVE EQUAL RIGHTS!

There shall be equal status in the bodies of state, in the courts and in the schools, for all national groups and races; All the people shall have equal rights . . . to develop their own folk cultures and customs; . . .

The position suggested appears to approximate Lijphart's corporate federalism and to lead to an argument for extensive decentralization. Later, however, the document is phrased in terms of individual rights.

The law shall guarantee to all their right to speak, to organise, to meet together, to publish, to preach, to worship and to educate their children.

This seeming contradiction within the document may perhaps be explained in the following manner. The context in which the Charter was drawn up was one of the denial of rights due to group identity, whether black, coloured or Asian. While the former quote indicates that these groups sought equal status with whites, the latter quote shows the actual intent underlying the document in respect of individual rights.[3]

Regardless of how one interprets the Charter, the ANC's position is clear. At the Mfuwe Lodge discussions, Tambo said that the ANC insists on one-person-one-vote in a unitary state, and Mbeki noted that the ANC rejected group structures and protections as they entrench ethnicity. These stances were subsequently emphasized in the Constitutional Guidelines, which bar politics organized on racial lines, call for a single, non-racial parliament, and support a Bill of Rights. The ANC specifically envisages a 'government, elected by, and accountable to the people, [and] a pluralistic multiparty state rather than the so-called democratic centralism of its one-party socialist allies . . .' (Adam 1988: 102). These proposals are laudable but, in the case of the attitude to race, it is a bit unsettling to note that it is of recent origin since it was only in 1985 that non-blacks were allowed on the ANC's National Executive Committee. Lodge (1983/4) suggests that the ANC's shift in this respect should, at least in part, be interpreted as a measure of its confidence regarding its internal support relative to the Black Consciousness Movement.

The commitment to a unitary state is consonant with the desire to reject group rights, but it is also due to the fact that the 'ANC sees federalism as a possible way of weakening the ability of a democratically-elected government to control the pace and direction of change' (Nidrie 1988b: 4). A sceptic might ask if the ANC's antagonism to federalism would be as strong if it could not expect an outright electoral victory, and could only hope for electoral majorities in the Eastern Cape and in Pretoria and Johannesburg – but it is probably churlish so to dispute the ANC's motives. For instance, and interestingly, the Constitutional Guidelines also lay great stress on the creation of a strong local government that would facilitate the access of the people to the corridors of power. While Wynand Malan (ex-NP leader of the New Democratic Movement, now also a member of the DP) disputes this, claiming that the ANC thinks in a 'prescriptive fashion and in concrete terms looks to a strong central government which will take almost all decisions' (*Weekly Mail*, 7–13 October 1988), on the face of it the Guidelines do not describe the machinations of a cynically power-hungry institution.

The Charter is less clear in its economic prescriptions, except that the democratic state envisaged is not the 'classical bourgeois democracy' with a limited public sector role in the economy (Hudson 1986: 8). For example, from the many relevant clauses in the Charter, public sector intervention is indicated by:

THE PEOPLE SHALL SHARE IN THE COUNTRY'S WEALTH!

The national wealth of our country, the heritage of all South Africans shall be restored to the people; The mineral wealth beneath the soil, the Banks and the monopoly industry shall be transferred to the ownership of the people as a whole; All other industry and trade shall be controlled to assist the well-being of the people; . .

THE LAND SHALL BE SHARED AMONG THOSE WHO WORK IT!

Restrictions of land ownership on a racial basis shall be ended, and all the land redivided amongst those who work it . . .

THERE SHALL BE WORK AND SECURITY!

All who work shall be free to form trade unions, to elect their officers and to make wage agreements with their employers; The state shall recognise the right and duty of all to work,

and to draw full unemployment benefits; . . . There shall be
a forty hour working week, a national minimum wage, paid
annual leave, and sick leave for all workers, and maternity
leave on full pay for all working mothers; . .

THERE SHALL BE HOUSES,
SECURITY AND COMFORT!

. . . Rent and prices shall be lowered, food plentiful and no one
shall go hungry; . . . Free medical care and hospitalisation shall
be provided . . . Slums shall be demolished, and new suburbs
built . . .

An interventionist state is indicated, but is there a suggestion
that the state should be socialist? Hudson's (1986: 32) view is
that 'Nothing in the Freedom Charter entails the elimination of
capitalism and the establishment of a transitional social formation
in South Africa. In fact the fundamental question from the point
of view of the transition to socialism of specifically working
class power is not addressed in the Freedom Charter.' Similarly,
Karis and Carter (1979: 63) write that the Charter omits 'any
reference to the abolition of classes and the establishment of public
ownership of the means of production. The nationalisation that was
proposed was, in the context of the Charter, characteristic of state
capitalism.'

Archer (1986: 20) contends that it is misleading to lay great
emphasis on the question of whether the Charter is a socialist or
a capitalist document.

the Freedom Charter is a point of departure not a blueprint
for system construction. Insofar as its economic causes entail
particular means, these are for a mixed formation with the
state's role quantitatively greater than at present, though by
how much remains unclear; and of course a rather different
mix of economic objectives is explicit and implicit in the
document.

In essence, the Charter provides a list of programmes without
prescriptions as to how they are to be achieved. As such, it
lends itself to neither a confirmation nor a denial of its system
presumptions. In short, as Archer (1986: 2) contends, the Charter
'is a programmatic statement couched in open-ended terms. *It is
not an operational blueprint grounded upon or compatible with only one
. . . form of economic organisation* (emphasis added).

It seems clear that the drafters of the Charter did not have a
clear idea regarding the means by which its goals were to be

achieved (Prior 1983/4; Innes 1985; Archer 1986). If one nevertheless wishes to reach some conclusion regarding the system presumptions implicit in the Charter, then a reading of its material goals leads to an interpretation of a welfare state (Prior 1983/4) based on state capitalism (Karis and Carter 1979). In its Constitutional Guidelines the ANC in fact calls for a mixed economy. The important point in respect of the Charter is that it is sufficiently vague to support diverse opinions and that continuing reference to it remains more a totemic necessity than an indication of an economic direction.

This ambiguity is no doubt a singular advantage to an alliance that contains divergent views and where it would be strategically dysfunctional to articulate clear positions (Padayachee 1986). The SACP views socialism as South Africa's future goal and the realization of the Charter – a national democracy having a supposed 'socialist orientation' – as the country's immediate goal. Ambiguity also facilitates changing interpretations over time, depending on internal struggles within the ANC. Is there reason to believe that appropriate interpretations of the Charter are mobile? Indeed, outside the SACP, what are the interpretations, not of academics, but of the actors themselves?

Mandela's view in 1964 was that the Freedom Charter

> is by no means a blueprint for a socialist State. It calls for redistribution, but not nationalisation, of land; it provides for a nationalisation of mines, banks and monopoly industry, because big monopolies are owned by one race only, and without such nationalisation racial domination would be perpetuated despite the spread of political power . . . nationalisation would take place in an economy based on private enterprise. The realisation of the Freedom Charter would open up fresh fields for a prosperous African population of all classes, including the middles class.
>
> (Cited in Karis and Carter 1977: 787)

Interestingly, Mandela argued for redistribution because capital is in the hands of whites; he eschews socialist motivations – 'nationalisation would take place in an economy based on private enterprise'. His primary concern appears to have been the creation of opportunities for blacks.

More recently, Tambo has said that 'the Freedom Charter is not formulated on the basis of any ideological positions. . . . The Freedom Charter does not even purport to want to destroy the capitalist system. All that the Freedom Charter does is to envisage a mixed economy . . .'[4]

Mandela and Tambo (the latter's views apparently are similar to those of Mandela) are ageing leaders and there is a distinct possibility that the ground is shifting from underneath them. Lodge suggests that this may, in fact, be what is happening. He writes that 'With the renaissance of popular political culture during the post-Soweto era there has developed a profound and widespread antipathy to capitalism' (cited in Sampson 1987: 162). This is also evident 'in the anti-capitalist polemic of virtually every black trade union spokesman . . .' Thus, one has to question whether, in the following quote, Adam and Moodley (1986: 198) are not referring more to the ANC 'line' than to popularly held views within the ANC when they assert that 'Black nationalism, on the whole, aims at capturing capitalism for its own benefit rather than overthrowing it.'

Observers have considerable difficulty in interpreting the socialist debate. There are two problems here. The first is that sources vary in their opinion, depending seemingly on their own commitment. The second is that it is entirely unclear what the actors mean by socialism.

In the first regard, a survey by Orkin (1986) is illustrative. Orkin presents broad support for socialism, based on the results of his survey of black South Africans, which revealed that 77 per cent favoured a socialist South Africa. He also writes that the

> ANC retains a commitment to socialism, which is shared not only by the non-racial UDF in its admiration for the Freedom Charter [thereby implying that the Charter is a socialist document], but also by the other pro-disinvestment radical groups and unions, and by Bishop Tutu, who has lately affirmed his broad belief in socialism on account of its egalitarian and humanitarian thrust.   (Orkin 1986: 58)

Following Orkin, the distinct impression is that it would be correct to view the ANC as a socialist organization, but there are two reasons for interpreting his views cautiously. One is simply that he contradicts Thabo Mbeki, a member of the ANC's National Executive Committee, who states that 'The ANC is not a socialist party. It has never pretended to be one, has never said it was and is not trying to be' (cited in Adam 1987: 8). The other relates to the phrasing of the relevant question in Orkin's survey.

Question 9. Suppose South Africa had the government of your choice. There are two main patterns how it should organize people's work, and the ownership of factories and buildings. Which view do you most support?

• the capitalist pattern, in which businesses are owned and run by private businessmen, for their own profit.
• the socialist pattern, in which workers have a say in the running of the businesses, and share in the ownership and profits.

If ever there was a question whose phrasing was intended to elicit a certain answer . . . !

Nevertheless, it would be glib, on these grounds alone, to dispute the increasing popularity of socialist thinking. Socialism is especially popular among organized labour and it is therefore appropriate to turn to Cosatu's views on the subject. Many might object to the emphasis on Cosatu rather than the UDF (or the MDM) but, following 'two years of sustained government action, the detention of more than 30,000 people and the death of a further 2,500 – most of them linked to the UDF – the dominant legal opposition force has been organisationally decimated' (Nidrie 1988a: 10). Moreover, the UDF has revealingly defined its constituency as consisting of academics, university students, school pupils, teachers, lawyers, doctors, clerics and the like (cited in Adam and Moodley 1986: 92) – such persons are unlikely to have the same interests as the working class. As stated by Freund (1986a: 124; emphasis added), 'The point here is not to drive a wedge between different classes in the black population which, undoubtedly, share oppressive conditions and deep feelings of national affinity, but rather to suggest that *only working class organisations politically conscious and distinctive, have an interest in pressing forward with socialist demands meaningfully*, . . .'

The *de facto* banning of the UDF has concentrated attention on Cosatu. In respect of the Congress's views, Saul (1986: 24) cites Jay Naidoo, the Congress's general secretary, as saying that Cosatu seeks 'a restructuring of society so that the wealth of the country would be shared among the people. . . . Cosatu was looking at alternatives which would ensure that any society that emerged would accurately reflect the interests of the working class'. This 'radicalization', employed by Saul with a view to furthering his own presentation of an increasingly socialist struggle within South Africa, may be contrasted with the views of Elijah Barayi, president of Cosatu. He states that 'Yes, I believe Cosatu is a socialist organisation and I would like to see a socialist state in South Africa. I speak of socialism as practised by the Labour Party in England' (cited in Adam 1987: 8).

This leads to the next issue: what do the various actors mean by socialism? Present-day Labour Party socialism does not mean state ownership of the means of production. Uncertainty regarding the meaning of socialism is suggested by a few more examples. Kenya's model of African socialism provides a well-known case of a capitalist path of development, and in Zimbabwe 'a "Marxist–Leninist" presides over a society with less claim to socialist structures than most countries in Western Europe' (Freund 1986a: 125). In other words, one should probably accept that most blacks view themselves as socialist, but one should then take this to mean majority rule, minimum wage legislation, a racial redistribution of public spending, and policies more conducive to the effective functioning of the unions.

In summary, it is evident that the question being addressed has shifted ground. I began with the desire to know about current interpretations of the Freedom Charter, but find that the question presumes a broader issue, namely, is the ANC a socialist organization? In attempting to answer this, it appears both that the official position of the predominant African nationalist faction is for a mixed economy and that socialism has a distinct appeal, both for many within the ANC and perhaps for greater numbers among its allies. One has always to remember that, in Tambo's words, the ANC is 'not a political party, it is a national movement and has within it people of all political persuasions'.[5] Thus, frustratingly, the Constitutional Guidelines call for a mixed economy but leave the nature of that economy unspecified. It is probably best to be content with Tambo's point at the Mfuwe Lodge discussions where 'He said he was often asked whether the new South Africa would be socialist or capitalist? It was not possible to say that the fight was for a capitalistic state or a communistic state – the fight was simply to be free.' Beyond that, decisions would be submitted to the people.

The ANC and the SACP

'The death of apartheid will merely be a milestone on the road to determining the future shape of South African society, and will not mark the end of the struggle' (Yudelman 1987: 250). Recognizing this, and with an alarmed view of the power of the SACP within the ANC, Lombard has stated that 'If an unqualified one-man-one-vote election was held today in the Republic a non-white leader with a communistic programme would probably attain an overall majority based on a pledge to confiscate and redistribute the property of the privileged classes' (cited in Zille 1983: 64). This view typifies white perceptions of the ANC and the nature of its relationship with the SACP. Is it accurate?

The South African government lays great stress on the significance of the ANC–SACP alliance. In Karis's (1986/7) view, known

SACP members form a small minority of the ANC National Executive Committee. Later informal discussions I have had suggest that SACP members may well form a majority of the National Executive. The membership of the SACP itself is small since the party 'considers itself an elitist, vanguard group' (Adam 1988: 97), but the role of the SACP within the ANC is important because white fears regarding SACP influence remain a major obstacle to potential negotiations. It is therefore worth pursuing the question whether, after gaining power, the ANC would promote the radical redistribution portrayed by Lombard.

On the one hand the SACP's notion of a two-stage revolution, described below, provides for a continued role for privately held capital after the democratic or first stage of the revolution. On the other, a reasonable perspective is ensured by the terms of the overlapping membership in the ANC and the SACP: 'Both the ANC and the Communist party recognize that the organizations are separate and independent from each other, that the ANC is the leader of the alliance, and that communists who are members of the ANC are obliged to be loyal to it' (Karis 1983/4: 396). Further,

> Tambo stated that the ANC will allow no deviation from the Freedom Charter and that the SACP had bound itself to that Freedom Charter. He said that the SACP presented no problems to their people as the ANC is a national movement which could embrace communists, moslems, christians and anybody else who wished to join. . . . Tambo said they were only allied to the SACP by means of their joint adherence to the Freedom Charter and overlapping membership.
>
> (Mfuwe Lodge discussions, 1985)

The SACP and ANC literature has it that South Africa will first undergo a revolution led by a united front of opponents to apartheid. Once a 'national democracy' has been established, socialist and working-class organizations can pursue a second transformation – the process by which this second transformation is to occur is left unspecified. The notion of two stages is implicit in Ramaphosa's[6] saying that the Freedom Charter represents the minimum demand of the National Union of Mineworkers (NUM). Indeed, he adds suggestively that 'the end of national oppression has to be achieved first' (Ramaphosa 1987: 50). The notion of a two-stage revolution is in distinct contrast to many socialists who believe that 'racial capitalism' is the source of oppression, and who argue that the struggle for socialism is, at the same time, a struggle for freedom from race and class oppression. The latter is, of course, somewhat ironic, socialist countries having a rather poor record on human rights.

With respect to white fears of a radical redistribution of property, it is also worth noting that considerable space is left for capitalist development. The space is made available for two reasons. First, the working class is considered to be too small to engage in a successful revolution. In this respect, 'The theory of national-democracy was developed by Soviet and allied Marxists in an effort to deal with problems confronting the struggle for socialism in societies in which the level of development of the productive forces is considered comparatively very low and in which the working class is both numerically and politically insignificant' (Hudson 1986: 6, 7).

Secondly, it would be a mistake to make the state responsible for the material welfare of the people when the productive capacity of the country is unequal to the task. Premature moves in this respect might well provoke a reaction against socialism. Thus,

> it would hardly be expedient to put a total ban on the development of private capital, even in countries which have already moved further than others along the path of social progress. The public sector is not yet able to guarantee a country the necessary goods. So great is these countries' backwardness that it is necessary to use all available resources, under state control, of course, for economic development. Total prohibition of private capital might also do political damage. The revolution is at a democratic stage. This would be a sectarian policy, which might result in the defeat of the progressive forces and ultimately in the victory of imperialism.
>
> (*Mirovaya Ekonomika Mezhdunarodniye Otnoshenia/ World Economy,International Relations*, 1964, 6: 349; cited in Hudson 1986: 17)

The SACP is not covert regarding its intentions. For example, Slovo, the head of the SACP, has stated that:

> The *strategic aim* of our party is to destroy the system of capitalist exploitation in South Africa, and to replace it with a socialist system in which the ownership of the means of production will be socialized and the whole economy organized to serve the interests of all the people. Such a society can only be achieved if political power is placed firmly in the hands of the working class. The *immediate aim* of the party is to win the objectives of the national democratic revolution.

. . . At the same time it is the duty of our party to spread its ideology of Marxism–Leninism.

> (Dubula, a pseudonym for Slovo, 1981; cited
> in Prior 1983/4: 190; emphasis in original)

Commitment to the two-stage theory probably represents a correct assessment of the strength of the working class in South Africa (Freund 1986a). Nevertheless, adherence to the theory and support of African nationalist groups belies experience because, as Freund has remarked, during the post-colonial period socialism has seldom emerged from its 'national democratic cocoon'. In this respect,

> It may well be the case that increased state power over production is a necessary preliminary to the eventual social appropriation of the means of production. A necessary condition of the latter is however that the proletariat and its allies exercise state power. Unless this is the case it is even arguable that the establishment of a national democracy and the pursuit of the 'non-capitalist' path of development are likely to facilitate the eventual consolidation of (an admittedly somewhat modified form of) capitalism in South Africa rather than enable a transition to socialism.   (Hudson 1986: 23)

Likewise, Saul (1986: 20) maintains that the '"most important" determination of whether the revolution will move toward the "true liberation"', which socialism represents, is precisely the 'role played by the working class in the alliance of class forces during the first stage of the continuing revolution'. The likely prominence of the working class in this respect is unclear. The ANC views the working class as '*a* leading but not *the* leading force' (Adam 1988: 114), and the SACP believes that the 'unions are merely another important pressure group' and sees itself as the sole representative of the working class as a whole (idem.). Morobe, an important figure within the UDF and the MDM, further reveals potential tensions when he writes that the MDM:

- is committed to the leadership of the working masses in the struggle for total liberation;
- accepts the African majority as the main force of the struggle; and
- recognises the ANC as the vanguard in the National Democratic Struggle . . .

As matters stand, the ANC calls for 'the constitutional entrenchment of the independence of the trade union movement' (Nidrie

1988b: 5). The advisability and likely success of this arrangement is something that is debated off the record. Suffice it to say that

> the precise nature of the relationship between the trade unions and the national liberation movement is one of the unresolved issues in the labour movement. Some trade union leaders would emphasize the autonomy of the trade union movement either because they believe workers should not be involved in politics or because they believe workers have the capacity to develop their own distinct working class politics and even working class party. Others would argue that no such distinction can be drawn and that trade unions should be closely linked to the national liberation movement.
>
> (Webster 1986: 21)

The shortcomings of two-stage theory have been pointed to above and persuasively argued in the papers by Hudson (1986) and Freund (1986a). Their intellectual severity, however, tends to distract one from what are good reasons for the alliance between the SACP and the ANC. From the ANC's viewpoint, it strengthens the array of forces aligned against apartheid and facilitates access to military supplies from the Soviet Union. From the point of view of the SACP, 'it is anxious to establish its credentials as part of the liberation movement, which it does through its links with the ANC' (Prior 1983/4: 190). A delay in the pursuit of socialism is also appropriate if it is a correct assessment that the black working class is still too small to enable a successful socialist revolution. This is especially the case if, as Slabbert and Welsh (1979) hold, the workers experience their oppression as primarily a racial phenomenon. Following the democratic revolution, it will clearly be considerably easier to present their situation as intrinsic to capitalism. In this situation, the SACP should engage in preliminary activity directed at enhancing its organizational capacity among the working class – all this being greatly facilitated by the legitimacy gained from its alliance with the ANC.

Capital and the ANC

The tragedy for the country remains the belief amongst most whites that to surrender political control is to commit cultural and economic suicide. Even so, the South African polity does not represent a coherent alliance between and within capital and branches of the state apparatus (Greenberg 1984; Lipton 1985;

Freund 1986b). In particular, capital 'has now become vitally concerned with the legitimation crisis of the state and the way that crisis is threatening the very future of the free-enterprise system and of capitalism' and has begun to doubt whether it 'can, in its most literal sense, live with the present state' (Yudelman 1987: 254, 264). O'Brien (1986) even suggests that liberal capital, convinced of the inevitability of majority rule, is increasingly prepared to deal with the ANC; that it will accept Mandela as the prime minister and Tambo as the minister of defence, provided that capitalism is retained.

There is, therefore, a degree of fragmentation in the response to the liberation struggle and to the ANC, and McCarthy (1986) has indicated that there is hope that South Africa will soon participate in a Lancaster House Zimbabwean-type settlement. While there have been some preliminary diplomatic skirmishes between the ANC and the government, this optimistic anticipation of the big event may be misplaced. Certainly, the ANC's view is extremely sceptical. It holds that NP talk of negotiations is intended to 'defuse the struggle by holding out false hopes of a just political settlement which the Pretoria regime has every intention to block' and to 'defeat the continuing campaign for comprehensive and mandatory sanctions'.[7]

My hope would be for a preparedness, not simply to undertake reforms in a racially structured society, but to relinquish ultimate control. Since the conventional wisdom confidently asserts that the war is unwinnable for the NP, and since many of us are equally confident that the repressive resources available to the government ensure that it cannot be overthrown, we end up with a stalemate and it would seem obvious that the next step is a Lancaster House agreement. It is hopeful, therefore, that Tambo has committed the ANC to negotiations (cited in Sarakinsky 1988), and it is encouraging that some NP members recognize the urgent need for negotiations. Nothnagel, a prominent NP Member of Parliament on the left of the party, has asserted that 'the majority of NP MPs accepted they would eventually have to negotiate with the ANC' (*Business Day*, 26 May 1988). However, in the words of an Afrikaner academic, Sampie Terreblanche, NP reforms are still 'little more than a public relations exercise to improve the image of apartheid and not endanger the Afrikaners' (cited in *Work in Progress*, 1988, 52: 49). In my view, NP reforms seldom involve genuine reform and are usually attempts to deceive and co-opt, while effectively simply revising the terms under which white supremacy is retained. While there are no saints on the South African stage, there are villains who are more dastardly than others.

Capital must be particularly concerned with how to prompt the government to create conditions amenable to negotiations, such as the removal of all racial legislation, one citizenry and the release of all political prisoners. The urgency of negotiation arises from the implications of the manner in which the next government comes to power. As noted earlier, if it occurs as a result of negotiation among relative equals, some compromise among the above options is likely, but, if it is the outcome of a victory in a civil war, less desirable outcomes seem certain. No doubt aware of this, Relly, head of the Anglo American Corporation, complained at the Mfuwe Lodge discussions that the NP listens but does not respond. This view, of a disjuncture between the leading representatives of capital and the NP, obviously represents a profound paradigm conflict between liberals and Marxists, and the reader's response is inevitably conditioned by his or her persuasion. Suffice it to assert that on the question of ultimate control of the state there are conflicting views. The neo-colonial experience of the last 30 years has made it clear to capital that direct control of the state apparatus is not a precondition for profits.

Capital has a specific motivation for dealing with the ANC, namely the feeling that socialism within the unions and the 'wider political culture' among blacks is ahead of the socialist influence within the ANC (Lodge cited in Sampson 1987: 162). One can also be quite certain that capital is alert to the possibility of strengthening African nationalism within the ANC and of ensuring, through this faction, that the post-apartheid mixed economy is weighted in favour of the market. (Prior, 1983/4, also points to the ANC's improved relations with the West creating tensions in the alliance.) The terms under which a deal could be struck are speculated on in discussion in South Africa and in different ways by Freund (1986a), Moll (1986a) and Adam and Moodley (1986), but it would be a mistake to believe that this could or should involve an attempt to cut the SACP out of the picture. Adam (1988: 120) makes the point that the SACP tends 'to be more pragmatic concerning negotiations and the post-apartheid economy than other forces in the alliance'.

African nationalists have little reason to accept formal democracy without associated economic concessions. Capitalism has not served the majority of South Africans well and I have little confidence that, in the future, an unattended capitalism would perform better. It is clear that 'Unless a significant social and economic restructuring occurs based on a developed notion of redistribution, only a small proportion of those who now feel oppressed would be able to benefit substantially' (Freund 1986a: 124).

Should the ANC, in seeking greater control of the economy, persist with its threats of further nationalization? The term 'further' is employed because, in 1988, the state owned 40 per cent of all the wealth-producing assets in South Africa (Riordan 1988). The state is now committed to a privatization programme – part of the process of moving economic power out of the hands of a future government. Of the balance of the wealth-producing assets, 40 per cent is owned by corporations quoted on the Johannesburg Stock Exchange and the remaining 20 per cent is owned by individuals and foreign corporations (ibid.) Since, in 1987, four corporations had a controlling interest in over 83 per cent of the companies listed on the Exchange, an extraordinary degree of concentration of control over the South African economy is indicated. The target for the government would, in large part, be those four corporations: the Anglo American Corporation (60.1%), Sanlam (10.7%), the SA Mutual (8.0%) and Rembrandt (4.3%) – with Anglo American being the foremost objective.

One problem with nationalization is that it simply transfers productive capacity and employment from one sector to another without creating additional productive capacity and jobs – the expenditure involved in nationalization has an inordinately high opportunity cost. Thus, when Riordan (1988) calculates the cost of acquiring the four corporations – perhaps up to R106,062 million – or, less ambitiously, a controlling interest – about R30,528 million – and compares this with the central government budget in 1987/8 of R46,318 million and a housing allocation of R781 million, he is led to conclude that government cannot afford to nationalize these corporations. There may be reasonable motivations for wanting to establish some control over monopoly industry, and also for extracting further revenue from that industry, but nationalization is not an effective means of achieving either goal.

The Freedom Charter was written in the 1950s, when nationalization and welfare states were popular and economically feasible in Europe. It is only since then that the consequences of nationalization have become well known. The retreat from public ownership of enterprises in Europe and in many socialist countries, and the failure of state-owned enterprises throughout Africa, should be instructive. For instance, the African experience is largely that nationalized industries do not increase the public revenue available to the state. The African parastatals are inefficient, excessively susceptible to pressure for maintaining unrealistically high employment and wage levels, consume comparatively higher levels of scarce capital, and demand management expertise that is already in short supply (International Bank for Reconstruction and Development 1984). In short, they are seldom profitable and usually require subsidies. Ironically, in terms of nationalization agreements, public subsidies might in part go to meeting the state's obligations to the former owner of the enterprise. Sklar (1975) showed in his analysis of the mining industry in Zambia that nationalization agreements were, in some cases, more lucrative for the former owner than ownership of the industry had been. The alternative – unfair nationalization (compensation not at the market value) or the seizure of enterprise – promotes capital flight and inhibits foreign aid and participation in the economy, all of which are essential if the post-apartheid economy is to thrive.

A response to the above is no doubt that, these problems notwithstanding, the Afrikaner-controlled state used employment in government and parastatal companies to alleviate unemployment amongst the 'poor white' Afrikaners, and that it is now the turn of blacks to use jobs in the public sector as a means of reducing unemployment. The point is acknowledged, and it will no doubt be a precondition to any form of agreement between capital and the ANC that demands of this sort are met. Nevertheless, Moll (1986a, 1986b) argues the need for alternative ways to influence the behaviour of South Africa's monopoly industry and suggests that there are other means, besides 'blunt nationalization', of doing so. In this respect, Rotberg[8] was prescient when he suggested that the leaders of the ANC may well be sympathetic to an approach that is not based on nationalization because, from their base in Lusaka, they have witnessed the protracted problems associated with Zambia's nationalized mining industry. Nidrie (1988b) reports that the ANC's Constitutional Guidelines are, indeed, hesitant in respect of nationalization, but he does not go as far as Rotberg in suggesting that the leaders of the ANC may, privately, be somewhat less than committed to nationalization. Southall (1986: 10) anticipates the likely outcome as an 'emulation of the Afrikaner 1920's and 1930's strategy of subjecting [the mining] sector to heavy taxation [in order] to underwrite the development effort elsewhere'.

I imagine that the ANC would have two immediate economic goals. One would be that there should be increased opportunities available to their constituency, in the form of public sector employment, strong unions, and soft loans and assistance to black entrepreneurs. The other would be to increase public sector influence over the private sector. To this end, if Southall's prediction is found to be politically unacceptable, my suggestion is that, in order to so influence capital, the state could seek control in a few strategically located corporations. In this respect the much decried, extremely monopolistic character of the South African economy would serve the new government well, for a couple of well-placed highly selective investments would enable wide-ranging influence while, at the same time, allowing the private sector to pursue growth, efficiency and profits (and thereby also to generate tax revenue).

The last issue concerns whether the ANC will accept some form of federalism. Both Lijphart (1985) and Adam and Moodley (1986) refer to occasions where ANC representatives have indicated a willingness to consider such a constitution. They might also hold that the ANC is not so much against federalism as it is against the homeland system. Mbeki, for instance, creates this impression when he says 'We would not want to federate these constitutional units that the apartheid system has created. We cannot say we need a federal structure which must recognise the reality of the Bantustans.'⁹ Nevertheless it is clear that these authors are grasping at statements that do not reflect current ANC policy. At the Mfuwe Lodge discussions Tambo was unequivocal about the ANC's desire for a unitary state and the ANC has plenty of reason to be suspicious of the federal option. This is not because federalism is antithetical to socialism. As has been noted, the ANC does not presume that the future is socialist and, anyway, there are federal socialist countries. There is also no reason to fear that federalism is 'incompatible with redistribution policies and steps towards social democracy or an extended welfare apparatus, which will be necessary' (Slabbert and Welsh 1979: 143). The reason for the suspicion is found in the extraordinarily open writings of Lombard and Du Pisanie (1985) as well as Louw and Kendall (1986), which appear to suggest that federalism's purpose is to restrict the powers available to a future majority government. Given such intentions, it would be more than unreasonable to expect the likely victor at the polls to concede a federal system.

Lijphart and the DP approach the issue from a much more considered point of view and, if they are correct that unitary governments exacerbate conflict in divided societies, then federal options are worth considering. The question, therefore, addresses the nature of South African pluralism. The government's notion of ten black nations and one white, one coloured and one Asian nation is rejected not only formally by the ANC, but also by a large majority of South Africans.

It must not be overlooked that South Africa is not, yet, a wealthy country and the resources available for economic growth, employment creation and welfare policies are limited. The post-apartheid era will be characterized by intense competition over the allocation and distribution of those resources. Even given the more constrained divisions Adam anticipates – English-speaking whites with most Indians, some coloureds and some blacks; Afrikaners with most coloureds; and most blacks with some coloureds and whites, but who themselves split among Zulu/Xhosa lines – the stage is set for an unremitting struggle for control of state policy. The federalist view here would be that a constitution that promoted power-sharing would contribute to stability and growth. In sum, though, one must concede that, to the extent that the initiative has passed to the ANC, the time for federalism has also passed. Whereas federalism might have been willingly accepted ten years ago, it is difficult to escape the conclusion that it has become passé.

Towards the post-apartheid state

An attempt at detail here would be both overly speculative and silly. Given Southall's (1986: 3) warning that 'the ideological knives are already being sharpened', it would also be a hazardous undertaking! No one can claim special insight into the outcome of the future negotiations. One has to keep the purpose of this book in mind and seek only as much detail as is required by that purpose. Unnecessary speculation on the colouring of the route-markers can only serve to detract the readers' attention from the subsequent policy debate – far better that I avoid such uncertain ground. I will therefore provide brief arguments regarding why I am led to the state I hypothesize.

In reference to South Africa's future economic system, one clearly needs a sense of the division of responsibilities for the production and distribution of wealth between the public and private sectors. It is immediately apparent, then, that I do not foresee pervasive public ownership of the means of production. One reason for this view concerns the means by which I anticipate that the ANC will come to power, my assumption being that there will be a process of negotiated change. If majority opinion, including that of the ANC, is correct that the NP cannot be overthrown by military means alone, then negotiations must follow. Moreover, the 'Negotiations [will] be mainly about the terms of ANC inclusion and the restoration of a new parliamentary democracy than about the transfer of power' (Adam 1987: 9). It further goes without saying that the likelihood of successful negotiations will be considerably enhanced if they are concerned more with addressing national liberation than with redressing class oppression.

The economic context of the negotiations will also play a forceful role in determining the outcome. As things stand, the world is increasingly faced with a 'go it alone' South Africa whose economy is winding down, with unemployment and poverty becoming critical. Indeed, it is this run-down economy that will contribute to the willingness on the part of whites to negotiate. Consequently, the ANC negotiators will have to confront the reality of a South African economy that is dependent on international support for recovery, so the stage will have been set for Western intervention. Thus, 'I foresee not the revolutionary overthrow of the state but its erosion from below and at some point the crucial intervention of Western governments to "save South Africa for capitalism"' (Southall 1986: 2). If one bears in mind that the economy needs about a 6 per cent per annum GDP growth rate if unemployment is to decline, current low (and often negative) per capita growth, projections of a feasible maximum 3.6 per cent growth rate (Blumenfeld 1986: 9) and the images Sunter (1987) evokes concerning possible growth rates of up to 10 per cent per annum if accommodation is reached with foreign investors, then it is clear that Western intervention may appear very attractive. Several participants at the York conference on the post-apartheid economy were at pains to emphasize South Africa's reliance on foreign capital and direct investment.

However, this does not mean that the post-apartheid government will embrace capitalism. It 'is rather more than likely that the post-apartheid government will espouse socialism in some guise (even if all it aims for or achieves in practice is a welfare-oriented capitalism)' (Adam 1987: 9). One reason for this claim is that the leadership and much of the support of the ANC is petit bourgeois. It is this class that will be the most immediate beneficiary of the transition to a non-racial state and it may see greater material gain in national liberation than in structural changes that might make that gain less possible. This 'sell-out' itself needs to be seen in the light of the fact that negotiations are underway and that an ostensible, as opposed to an effective, transition to socialism would either wreck the negotiations or would so threaten the prospects for future growth that concessions in favour of a mixed economy would seem advisable.

This prediction, of only the visage and not the substance of socialism, reflects the typical experience of the subordination of the trade union movement to the national liberation movement. However, it ignores a number of imponderables, including the comparatively greater strength of the unions and the level of industrialization in South Africa. Why, then, be sceptical about the future role of the unions? Are they not central to even the ANC's pursuance of the struggle?

My view is that their prominence partially derives from the lack of political rights for blacks. It seems to me that with the current vacuum resulting from the illegality of the ANC and the banning of other actors, the relative power of the unions is exaggerated. Once such rights are guaranteed, in the absence of a working-class party their centrality diminishes. The critical factor is whether organized labour will form a part of the first, future, democratic government. Although the strength of the ANC is founded on the support of the urban labour force, the ANC apparently specifically does not want to see organized labour 'tied in to the ruling party' (Skweyiya, cited by Nidrie 1988b: 5) – the unions are to remain independent. Thus, while Cosatu exists as an independent union movement, it both calls for 'the leadership of the working class in the struggle' (*South African Labour Bulletin*, 1987: 54) and acknowledges the ANC's being the 'overall leader' (Saul 1986). As noted, the potential for tensions in the relationship between Cosatu and the ANC are obvious.

The possibility of forming a working-class party is itself problematic given that, by the year 2000, formal employment will engage only about 45 per cent of the black labour force and also that, as late as in 1983, only 9 per cent of that labour force was unionized (Maree 1986). A post-apartheid government could easily claim that it represents the nation and that Cosatu's call for 'a society where the interests of the workers will be paramount' (*South African Labour Bulletin*, 1987: 54) is selfish. In this regard Cosatu's claim that it is 'the home for every worker in South Africa – employed and unemployed' (idem.) leaves one a bit incredulous.

Conflict with the unions might also derive from the fact that the government's fiscal capacity is 'crucially dependent upon the profits of the gold mines' (Blumenfeld 1986: 26). When higher wages for mineworkers are set against reduced education or housing expenditure, higher wages for what will increasingly be viewed as a privileged stratum of the labour force may seem untenable. Put simply, income transfers between capital and labour within the modern sector do not reach the unemployed and the poor (Moll 1986b), except perhaps minimally through transactions with the informal sector.

Wage restraint will also be suggested by inflationary fears and by the implications of inflation for those among the non-union population who are living in poverty. In addition, given the 'militancy and relatively low productivity of the South African workforce' (Freund 1986b: 13), wage restraint is necessary if South Africa is to be competitive within the international division of labour and if it is to attract foreign investment. In particular, Bell (1983, 1987a) has shown that to dispute the discipline of the international marketplace is to misunderstand the competitive pressures facing South African industry. In the words of Grosz, the former prime minister of Hungary, 'The way I see it, if a worker's performance isn't a plus, he should not get even a 1 per cent increase in pay. If workers don't like that tough talk, the world market isn't the least bit interested' (*Newsweek*, 18 July 1988: 23).

Within this environment, likely economic policies will include legislation that favours the unions; perhaps some dispersed nationalization, but not decisive action against the mines and major industrial groups; legislation and taxes intended to reap greater benefit from these economic sectors; the rapid growth and transfer of employment in the civil service to favour blacks; probably some desultory redistribution of marginal farms; and so on. There are a number of such formulations, but what they represent is the aim of a welfare state without the resources of a post-industrial economy. Conflict over the allocation of resources will be pervasive.

Lastly, with regard to the constitution, federalist attempts to weaken the future government, and thereby to entrench capitalism, are unlikely to succeed. One reason is simply that the ANC and the unions are alert to this artifice, but there are other, equally fundamental, reasons. Some images regarding South Africa's future consist of township youths leading the country to Khmer Rouge outcomes; or the Lebanonization of the country, a setting greatly contributed to by the battles between the UDF/Cosatu and Inkatha outside Pietermaritzburg. Thus, when Joe Slovo, head of the SACP, states on Zimbabwean television that, in the post-apartheid society, the ANC would not allow policies to be made in the streets, one can come to understand 'the likelihood of the major propertied interests increasingly coming to favour a strong state . . . to restore order in the townships and to keep the masses in control' (Southall 1986: 12). Once the need for majority rule is accepted, the ANC is likely to come to be viewed as the desirable purveyor of power because, as noted, socialism appears to be more pervasive in the unions and among blacks, generally, than it is within the ANC.

In other words, as was the case with a strong white state, so, too, a strong black state may serve capital's interests. It would engage in the policing of dissent and be necessary in order to confront the problem that social needs simply exceed the resources available to the state. In this light, a strong state both extracts from capital a deal for its major constituencies and restrains the demands expressed by other groups. What is more, in the interests of growth, which, more than any economic outcome, will ultimately best serve the interests of the masses, a unitary black government led by the ANC might indeed best facilitate capital accumulation.

Notes

1   This chapter is a revised and extended form of a paper in the *African Studies Review*, 31, 2: 35–60.
2   The Mfuwe Lodge discussions took place between the ANC and leading South African businessmen on 13 September 1985. Present were Oliver Tambo, Thabo Mbeki, Chris Hani, Mac Maharaj, Palo Jordan and James Stuart for the ANC; and Gavin Relly, Tony Bloom, Zach de Beer, Tertius Myburgh, Harold Pakendorff, Peter Sorour and Hugh Murray, as leading representatives of liberal capital. President Kenneth Kaunda of Zambia chaired the proceedings. References to the discussion are from summary notes by Tony Bloom.
3   The comments of an anonymous reviewer for the *African Studies Review* were especially helpful here.
4   The quote is from Mr O. Tambo's evidence to the Foreign Affairs Committee of the House of Commons, *Minutes of Evidence*, 29 October 1985, p. 5.
5   ibid.
6   Ramaphosa is the general secretary of the National Union of Mineworkers, the country's most powerful union and a member of Cosatu.
7   Statement of the National Executive Committee of the African National Congress on the Question of Negotiations, 9 October 1987.
8   In conversation.
9   The quote is from Mr T. Mbeki's evidence to the Foreign Affairs Committee of the House of Commons, *Minutes of Evidence*, 29 October 1985, p. 8.

Postscript

Early in 1990 the South African government introduced unexpected and welcome changes when it unbanned the ANC, the SACP, the PAC and other organizations, and released a number of political prisoners, most notably Nelson Mandela. De Klerk, the new State President, portrays the changes thus: 'there was no conversion on the road to Damascus, just a certain ripening, and an understanding that if we continued as we were, we were in a cul-de-sac' (*Business Day*, 2 March 1990). While the ANC were effectively beaten in military terms, a negotiated settlement with them had become a precondition to economic revival. De Klerk has therefore called for a political normalization which will lead to negotiations over future constitutional and economic arrangements.

In Chapter 3 the argument was advanced that the NP's constitutional position was unacceptable because it employed racially defined groups as the basis for corporate federalism. The NP now appears willing to accept that 'group structures' are unfeasible and instead calls for 'minority protections'. The ANC, on the other hand, argues for individual rights, but accepts that this will embrace rights in respect of language, religion and culture. A willingness to compromise is apparent.

Now that the NP has begun to adopt a more reasonable negotiating stance, the major differences between the NP and the ANC centre on a territorial federal versus a unitary constitution, future economic arrangements, and how to proceed with negotiations. The arguments surrounding federal and unitary systems are little changed from those described in Chapter 3 and will not be repeated here.

In respect of the economic arrangements, however, certain subtleties are emerging. The government was previously described as oriented to a free market environment. The issue appears to be more complex. It is a commonplace that the apartheid government wanted to privatize the provision of many services so that the struggle over their supply was not at the same time a struggle for control over the state apparatus. Now, however, while arguing that in a number of instances privatization is more effective in ensuring the supply of services, the government accepts the need to markedly increase spending in the area of social services and infrastructure, with a related decline of such spending in presently white areas. In addition, prominent figures within the government establishment are debating the nature of the structural changes in the economy that are needed in order to meet both black aspirations and ensure future growth. At the same time, a COSATU official publicly explores socialist economic arrangements that fall short of wholesale nationalization, which he refers to as a 'red herring'. No doubt influenced by events in Eastern Europe, even Joe Slovo of the SACP acknowledges that nationalization is not necessarily the answer for South Africa. He, like Tambo and the ANC's Constitutional Guidelines, asserts that issues of this sort must be the subject of negotiation.

These views regarding nationalization are interesting since they contradict Mandela who, on his release from prison, referred back to the Freedom Charter and demanded nationalization of the banks and mines. Indeed, it is ironic that the unions, who deem themselves socialist, explore alternatives while Mandela, who is not a socialist but is concerned with black economic advance, views this as a necessity. Or so it seems. It has been held that, in advancing these views, Mandela has strategic motives in that he wants to demonstrate a position which shows that he did not compromise with the NP while in prison (hence also the call for an accentuated armed struggle), and that he is attempting to create a bargaining chip which can be played against concessions from government and capital. The differences between the government and the ANC about how to proceed into the future are consequential. Mandela argues that those represented at the negotiating table should be elected, for otherwise they would lack a mandate. The government wants to deal with the leaders of organizations wishing to participate in negotiations. The former approach would considerably diminish the position of minority whites.

Ultimately, one has to accept that the political situation in South Africa is extremely fluid. This is evident also in speculation regarding the institutional integrity of the ANC. It represents a liberation movement, a coalition of groups, which is now trying to behave with the purpose and unity of a political party. This is itself a difficult task, but made more so by competition with the PAC, and by organizations such as the unions and the UDF which, while united with the ANC, are now less obliged to paper over differences. The ANC's image is presently enhanced; its future standing is likely to be diminished due to political tensions.

The events and views described above do not change the political–economic assumptions underlying the policies discussed in this book. Instead, the likelihood of negotiations and majority rule suggests that a book addressing options in respect of urbanization policy is timeous. Indeed, issues such as the lack of housing, access to land, rents and the cost of services are leading to mass protest throughout South Africa. Thus, at his welcome-home address in Soweto, Mandela drew a roar from the crowd when he decried the quality of housing available in Soweto. But not only are those lacking such housing unable to pay for it, the government itself cannot afford to provide housing of this standard on a scale that is sufficient to overcome the shortage. Stark policy choices lie ahead and debate concerning what is both desirable and feasible is now urgent.

4 *Land for housing the poor on the PWV*

Under apartheid the government has attempted either to prevent or to displace black urbanization, with the result that a democratic government will inherit a tremendous housing shortage. In this chapter I detail the main features of the housing problem, evaluate policy alternatives and call for a number of policies that could ameliorate the situation. This brave intent, however, is confronted by the enormous complexity of the housing problem in South Africa and it has therefore proved advisable to adopt a specific focus.

The supply of land for housing is an attractive issue because:

- the large majority of those requiring a dwelling unit are unable to afford a house – the best they can hope for is a serviced site, or participation in a squatter upgrading system;
- a progressive government can employ the location of urban development in order to overcome the disastrous legacy of the form of the apartheid city described in Chapter 2; and
- the form of many cities is likely to exacerbate the future demand for serviced sites. This is because a large proportion of, for example, the PWV's black population living in squatter settlements is poorly located on the borders of Bophuthatswana and KwaNdebele. This was the result both of influx control, and also of the fact that minimally serviced sites within the homelands were more easily available and cheaper than the extremely limited supply of formal housing and serviced sites in the urban areas. With a decent government and policies intended to increase the supply and to reduce the cost of urban land, many might choose to move into the cities.

The alternative scheme, squatter upgrading, is likely to be a subordinate policy in the PWV. In fact, a relative emphasis on squatter upgrading or land parallels regional differences in the form of urbanization. At Durban, for example, the borders of KwaZulu are close to the city itself and there are supposedly already about 2 million squatters living within the greater Durban area. Here, clearly, the primary issue is one of squatter upgrading. In the PWV, as I have said, the issue is not so clear-cut and, in selecting a focus on land, I am, to some extent, also choosing a regional focus. The other way of looking at this issue is to say that, if a regional focus is implied by the problem one addresses, then the PWV is worth selecting because it is the largest urban region, will be the destination of most migration and, consequently, it is here where future housing demand will be most pronounced.

The following material is directed at policies intended to increase the supply of serviced land on the PWV. The position I develop is that, besides acting so as to enhance the private supply of land, the government must itself undertake to supply serviced, well-located, often subsidized sites to the poor.

It is already apparent that I have ruled out simply the public supply of housing. While this might be readily accepted overseas, in South Africa there is still strong feeling that public housing is an appropriate response to the housing problem – indeed, it is specified in the Freedom Charter. I have therefore included two sets of arguments that support my position. One comprises a fairly technical evaluation of the history of housing policy since the 1950s and an assessment of what has and has not worked. Public housing, for example, will be seen to have been widely applied and to have almost invariably failed both to provide a sufficient number of dwelling units and to reach the urban poor. Self-help projects, a response to the failure of public housing, are now also not escaping criticism and the tendency is to look to a much greater private contribution to the supply of housing. Privatization, commonly in the form of public–private partnerships, is in many respects an important response as it provides more resources for solving the problem: capital and management expertise. Nevertheless, as I have argued, since the housing needs of the poor do not constitute effective demand in the market, they are not helped by privatization schemes.

My second set of arguments approaches the housing problem from a more political economic stance: I assess housing policy in socialist countries and show that there, too, the record is rather dreary. In my search for a better understanding of the conditions under which governments really do serve the poor I arrive at what seem fairly self-evident requirements: namely, stiff electoral competition and a sufficient number of the urban poor to make them an important constituency. In other words, the issue is not so much one of capitalism or socialism as of democratic government.

If one expects that South Africa's future is described by such conditions, then the overriding housing issue for the poor on the PWV becomes access to well-located, affordable land. There are many possible policies that may be employed to this end and I have chosen to deal with two particularly common alternatives. One involves indirect public intervention in the market for land, which is a likely response of a government operating within a mixed economy. Indirect interventions are not always very effective, however, and a government committed to meeting the housing needs of the poor would also be drawn to a more active role. The second alternative therefore commonly involves any of a variety of policies of direct intervention implemented by a land development corporation (LandCorp). Yet, while LandCorps may represent potent actors in the market for land, they are given to numerous inefficiencies and, it turns out, also promote inequitable outcomes.

My recommendations have been disciplined by the experience of housing policy elsewhere and, so, instead of my initial enthusiasm for substantial, wide-ranging interventions when I started this book, I come forward now with rather limited policy recommendations. Instead of grand plans, a set of carefully tailored policies appears all that is feasible.

The housing problem

Reports in the press have it that, in order to overcome the housing backlog, South Africa needs to build 1,000 houses a day, this despite the economy's supposed maximum and unrealized capacity of producing 70,000 houses a year. It is also widely held that the majority of blacks cannot afford a formal house, let alone a serviced site. The discrepancy between housing need, economic capacity and effective demand has produced a variety of different responses, for example:

- engineers comment on the shortage of artisans;
- construction companies lament the limited extent to which those in need of housing can afford housing;
- the housing finance industry defends itself against its unwillingness to provide bonds for the poor and complains of higher transaction costs and greater risk when lending to the poor; and
- sociologists at my university blame capitalism, as do many in the unions.

In the following sections on the housing backlog and housing affordability I attempt to introduce some rigour into the argument.

The housing backlog

Calculations of the housing backlog have very limited value. They generally start with an assumption that, in the absence of effective demand, every household, however it is defined, is entitled to a house. This method ignores or disqualifies many of the present housing arrangements such as squatter shanties, backyard shacks and families living in rented rooms, and proceeds to estimates òf housing requirements that neither the supposed occupants nor the government can afford. De Vos' (1987) estimate of the housing backlog, for example, is based on an uncertain population estimate. He then divides this estimate by an average family size of 5.94 and subtracts the existing housing stock from the result. The backlog so calculated draws considerable attention but is, in fact, not very meaningful. For illustrative purposes only, the backlog suggested for the PWV in 1985 was 350,000 dwelling units.[1] The backlog figure for all South Africa has been reported as about 1.8 million units (*Weekly Mail*, 21–28 October 1988).

These backlogs can be taken as the starting point for an already artificial estimate of land requirements. The next step involves determining the growth in the demand for housing by dividing the projected increase in population by an assumed average family size. In this case I divided Simkins' projection of the increase in the black population of the PWV of 4,927,367 by 6.1 (the average family size figure with which I am familiar); the result is just under 808,000 units between 1985 and 2000. As a matter of interest, the growth in demand for South Africa (excluding the independent homelands) will apparently be an average 127,000 units per annum for the next 20 years (*Weekly Mail*, 6–12 October 1989).

In order to determine the land required, one has then to assume an average population density. The ratio used by the Department of Constitutional Development and Planning is 15 dwelling units, or 92 persons, per hectare. Using this formula, the backlog on the PWV requires about 23,333 hectares and the black population growth in the PWV till 2000 requires another 53,558 hectares, producing a total need of 76,891 hectares of land. In 1985 and 1988 together, the government allocated 42,000 hectares for black housing, but at least 10,000 of these hectares are geologically unsuitable for housing.

In addition to the discrepancy between supply and demand, these figures refer only to land for housing and are themselves misleading. The land used for purely residential purposes in a city constitutes about 60 per cent of a city's area and the figures would have to be adjusted accordingly. However the adjusted figure would overstate the land required since much of the demand for housing will be expressed in the form of greater residential density. Urban residential density has two components – built density and the density at which the buildings are occupied. 'White' areas of Johannesburg have a low density in both respects, with the result that the city has a population density of about 1,245 persons per km^2 – below that of Los Angeles (Senior 1984). In comparison, densities of Third World cities range from about 3,000–4,000 persons per km^2 to 9,000 persons per km^2. Johannesburg can clearly sustain an increased population density and the existing infrastructure network of the Witwatersrand can apparently cope with approximately 1.4 million more residents.[2] This fact contributes to one's uncertainty regarding the housing and land requirements, for much of the backlog will be met by higher density occupation of dwellings, families sharing dwellings, renting garages and building backyard shanties – all outcomes that are common in middle-income countries and that usefully relieve the housing burden.

In sum, one really has no precise idea regarding the amount of land and housing necessary for the PWV and, although certain estimates have been put forward, they should be taken as providing only an order of magnitude.

Affordability

Table 4.1 shows the limited capacity of most blacks to afford housing. De Vos (1986) employed the Household Subsistence Level

Table 4.1 The ability of black households to pay for housing, 1985

| | Income group per month before tax (R) | | | | | | | |
	1–99	100–199	200–299	300–399	400–499	500–699	700–999	1,000+
% households	22.5	16.8	17.1	10.2	9.6	12.1	7.5	4.2
Cumulative %	22.5	39.3	56.4	66.6	76.2	88.3	95.8	100.0
Disposable income for accommodation[1]	—	—	—	41	141	341	641	
Affordable housing loan[2]	—	—	—	2,852	9,807	23,717	44,582	

[1] After deducting the HSL.
[2] Based on the amortization of a 25-year loan at 13.5 per cent per annum with equal monthly payments.
Source: De Vos (1987), Table 2.2.

(HSL)[3] as a measure of affordability when compiling the table. The HSL refers to the cost for a household of maintaining, over the short term, 'a defined minimum level of health and decency', and in 1985 the measure for a black household stood at R313 per month, excluding accommodation. It is clear from the table that, in 1985, 67 per cent of the country's black population could not afford a serviced site, which nowadays costs about R6,000 (including the land), and that approximately 80 per cent could not afford a minimal formal house costing around R20,000 (including the land).

The comparable 1988 figures for Johannesburg's black population are that 30 per cent could not afford a serviced site and more than 50 per cent could not afford a house (De Vos 1987). Their poverty notwithstanding, blacks living in the Johannesburg metropolitan area are better off than those living elsewhere in the country. However, the more important comparison is probably with the inhabitants of the KwaNdebele homeland, since it is here, along with other homelands, where an urbanizing black population was 'bottled up' during apartheid.

The incomes of KwaNdebele's inhabitants are shown in Table 4.2. These incomes are comparable, not so much to those in Johannesburg, but rather to the rest of South Africa. For instance, 56.4 per cent of South Africa's black population had a household income of less than R300 per month in 1985 and 61.1 per cent of KwaNdebele's population had a household income of less than R3,400 per year (R283 per month) in 1984. Contrary to what one would expect, proximity to the PWV labour market does not ensure higher incomes.

Since most of KwaNdebele's inhabitants are unable to afford even a R6,000 site, they are unable to acquire their own accommodation in the PWV. However, KwaNdebele's residents may not

Table 4.2 Income distribution in Kwa-Ndebele, 1984

Annual income (R)	% of households
<800	12.0
800–1,599	21.4
1,600–2,599	14.6
2,600–3,399	13.1
3,400–4,399	14.6
4,400–5,199	5.2
>5,200	18.1

Source: Bureau of Market Research (1984).

want to move closer to the cities since a location in KwaNdebele minimizes housing costs; for instance, in 1984 only 4 per cent of total household cash expenditure went to housing (McCaul 1987). The significance of this point probably hinges on changes in transport costs. Transport expenditure constitutes 7 per cent of household expenditure (but 17.5 per cent of a commuter's income) (McCaul 1987). It has already been noted that the government subsidy for labourers commuting from KwaNdebele is about R1,600 per labourer, but the government has stated its intention of doing away with the subsidy. Since most commuters cannot afford an urban serviced site, they are left with a choice between illegal squatting, renting rooms and tremendous overcrowding, building backyard shacks, or remaining in KwaNdebele with a vast reduction in income. In effect, 'orderly urbanization' in the PWV and the attractiveness of cheap housing in KwaNdebele are dependent on subsidized transport. At enormous social cost, the people are engaged in a balancing act between location, transport and housing cost.

A final and important equivocation is that private financial institutions have not serviced more than the top 20 per cent of the black urban income-earners. The majority of blacks obtaining private finance are those who qualify for government subsidy or who obtain support from employers. This conclusion, however, should not be understood in a one-sided manner as it is commonly the case that the urban poor do not benefit from formal finance. For example, Peattie (1987: 270) remarks that

> We also have relatively little good data on how building in the informal sector is financed, beyond a pretty clear understanding that banks do not do it. Family transfers and personal savings seem major sources. We do not even have factual information – and this is more consequential for policy – on how important a bottleneck financing is, compared to availability of land and of low-cost construction materials. Our discussions produced a good deal of anecdotal evidence of *reluctance* to engage in long-term borrowing.

Thus, in South Africa, one is not simply talking about the private sector's unwillingness to risk profits. Hardie *et al.* (1987: 35) have concluded that the 'principles of long term finance were poorly understood, and even among those committed to long term loans, the grasp of specific elements of mortgage finance was superficial'. Many expressed an unwillingness to consider loans.

The majority of blacks, in the PWV and elsewhere, can neither afford formal housing nor obtain access to bonds. Estimates regarding future urbanization and unemployment levels point to a worsening situation. The issue for the poor is not one of access to housing – this costs too much – it is one of access to serviced land.

The evolution of housing policy

In this section I comment on the history of housing policy in order to gather lessons for housing policy in South Africa. The history begins around 1950, when countries in Africa and much of Asia were gaining their independence.

Public housing

During the 1950s and much of the 1960s, government attempts to address housing shortages took the form of public sector construction. This was a mimicking in colonial countries of practice in the metropolitan countries. Even so, attention to the housing problem was not as pronounced as it is today. For one thing, the colonial authorities felt little pressure to prioritize the problem, partly because the extent of the urbanization process getting under way was not yet realized and the size of the forthcoming housing demand and pressure on services was still unanticipated. A related reason was that during this period the attention of the World Bank, which was to become a central actor in self-help housing programmes, was directed at national and sectoral economic objectives, not at the growth of cities and the welfare of the poor (International Bank for Reconstruction and Development 1975).

Public sector housing efforts are generally considered to have failed: not enough houses were constructed; they were constructed to too high a standard; they were too expensive; and they were usually allocated unfairly, with the most common beneficiaries being civil servants. Where the relatively poor did receive housing, 'downward raiding'[4] by the better-off, who themselves were inadequately housed, rapidly saw to it that the housing was reallocated.

There have been a few exceptions to the negative assessment of public sector housing, however. Singapore, in particular, draws repeated mention, even though its success in the provision of public housing has been explained as being due to a unique set of circumstances, which include a rapidly growing economic base (per capita GNP growth of 7.4 per cent between 1960 and 1978), slow immigration and a colonial legacy of extensive public land holdings (Hardoy and Sattherwaite 1981). To this list one can add a competent bureaucracy and the ready availability of funds, owing to the channelling of pensions and other savings into the housing sector. Despite these favourable conditions, housing in Singapore is in the form of high-rise apartments and, initially, families were allotted a single room.

Criticism of public sector housing is not only made in terms of its numerical failure. It was, in addition, often badly located on the fringes of the cities where large tracts of land were available. Location is critical for the poor since they struggle to afford commuting costs. Also, if they are dependent on occasional labour, they have to live close to where employment might be available. It is the relatively better-off who can afford to live some distance from their work.

The provision of a complete dwelling unit is also usually mis-matched to the poor's ability to pay for it and to their investment priorities. In a survey in Kenya, for example, the poor represented their priorities as follows: (1) food, (2) squatter housing, (3) school fees, . . . (9) standard house.[5] One should not presume an ability to speak for the poor, nor an ability to predict their behaviour. This is especially the case when, as is likely with those below the bread-line, 70–85 per cent of their income may be spent on food (Lipton 1983). Given such figures it is easy to understand why, when a few really poor persons were allocated sites in a site-and-service scheme in Kenya, their response was to sell or rent the sites and to set up illegal shanties elsewhere (Harris 1972).

A further criticism of complete units is that they do not allow the poor any flexibility in the allocation of their resources when coping with changing priorities – where household income is inconstant and where sudden illness can demand all one's income, fixed payment schedules can result in eviction. Complete units, moreover, are static, whereas household's needs are dynamic. When one is dependent on a minimal income, access to opportunity is more important than a two-roomed unit and a poor location. On the other hand, if one has a steady income and is supporting children and a grandmother, the inability to add an extra room may make the unit equally unsatisfactory.

Since the imagery contained within the Freedom Charter is largely that of public housing, it is interesting to see how its suggestions stand up against these criticisms.

THERE SHALL BE HOUSES, SECURITY AND COMFORT!

All people shall have the right to live where they choose, to be decently housed, and to bring up their families in comfort and security;

Unused housing space to be made available to the people;
Rent and prices shall be lowered, food plentiful and no one
shall go hungry; . . .
Slums shall be demolished, and new suburbs built where all
have transport, roads, lighting, playing fields, creches and
social centres; . . .
Fenced locations and ghettos shall be abolished, and laws
which break up families shall be repealed.

The Charter clearly reflects the period during which it was drafted –
at the time the Johannesburg City Council was beginning a massive
housing construction campaign in Soweto. The 1950s, in general,
were a part of the failed public housing era. Today, however,
the Charter seems rather distant from current understanding of
the housing problem. For example, one might observe that, since
lowering rents reduces the resources available to the government,
to do so is to favour those housed over those needing housing.
Similarly, were the public sector to provide housing, fiscal constraints would ensure that an insufficient number of houses would
be constructed. In the resulting allocative scramble, it would be
naive to think that the poor would benefit.

It remains to be seen whether the trade-off between quantity
and quality will be revealed in position statements by the African National Congress, United Democratic Front, the Congress
of South African Trade Unions and the National Union of
Mineworkers. NUM, at least, has begun to consider the issue.
A cynic in America predicted that these progressive organizations
would in fact 'fudge' the issue. She thought that an ANC-led
government would remain true to the Freedom Charter for about
five years, that is, for as long as it took to house major groups
of supporters. Thereafter, the cost of formal housing would be
'discovered' and site-and-service schemes would follow. Lest this
prediction be viewed as unduly cynical, given that now everyone
knows of the poor record of public housing, one has only to look
at housing policy in Zimbabwe. The government has, with one
exception, aimed to eradicate squatter settlements and has favoured
formal housing. The government has also set standards for housing
whose cost excludes the majority of the population and implements
full cost recovery for housing projects (Butcher 1986).

Self-help housing

The criticisms of public housing were drawn from observations
of self-help housing processes. Self-help housing, albeit frequently
without access to basic services, was viewed as providing precisely

those attributes that formal sector housing lacked. Above all, self-help housing reduced the cost of increasing the supply of housing.

THE COST OF SELF-HELP HOUSING:
Households can employ a number of methods of reducing the cost of a house. One is avoiding the land market, through either illegal occupation of land, or the occupation of public sector land or land unfit for any development, such as where it is subject to flooding. Illegal occupation certainly is the popular conception of self-help housing, no doubt because it is so common in Lima, where Mangin (1967) and Turner (1968, 1969) were located, and which they used as an example in their seminal studies. In practice, though, land is in most instances purchased by those who occupy it (Peattie 1987).

Low cost can also be assured through the partial use of one's own labour and through the use of building materials, such as mud bricks, that are not obtained from the formal sector. Peattie (1987) reports that, while the initial shack may be put up by the owner, there is usually increasing or total use of subcontractors as the building is consolidated. The frequent use of informal materials is also important since individual, low-volume, customers pay high prices to the formal sector.

The absence of finance charges and indirect costs – Connolly (1982: 158,9) estimates that 'taxes, professional fees and "promotional costs" may account for up to 24% of the total construction for government housing projects' – are another way in which costs can be avoided. Interest charges are reduced since additional construction is undertaken only when the inhabitant has the means to pay for it and, while relatively expensive informal loans may be employed, they are for small amounts – a new room or roof – rather than for the house itself.

Connolly (1982) also holds that the cost of informal housing is kept down through the acceptance of poor-quality housing. In so far as size is concerned, the validity of this point appears to be contingent on when the shantytown is surveyed. The amount of living space in a well-established shantytown may exceed that made available in formal housing projects, where one means of holding down costs consists of skimping on living space (Martin 1982; Peattie 1987).

SELF-HELP POLICY
The demonstrated successes of the poor were celebrated at the 1976 United Nations Conference on Human Settlements held at Vancouver. This conference was a landmark in the development

of housing policy because its recommendations endorsed site-and-service and squatter upgrading schemes. Self-help housing was to be encased in supportive government policies from the national to the local level. The private sector was barely mentioned. The government, it seemed, really could help.

Turner (1968, 1969) was especially prominent in attempting to specify how the government's role in the supply of housing should be redefined. He called for:

- doing away with restrictive minimum standards which, if met, make the house unaffordable for the poor;
- facilitating the supply of land, technology and credit to the poor;
- the acceptance of informal sector activities as a necessary and desirable part of housing in shantytowns.
- exploring ways of providing the necessary urban services as cheaply as possible, usually through pit latrines and shared water taps; and
- legalizing uncertain land tenure arrangements.

(While the poor are widely held to be prepared to construct only a rudimentary shack until their tenure is legalized, the feeling has recently emerged that legal tenure, which brings with it taxes and downward raiding, may cause the poorest among the squatters to have to move elsewhere – Doebele 1987.)

It would be naive, however, to believe that governments would readily adapt their housing policies to the recommendations above. It should also be recognized that government participation in self-help housing is not a new phenomenon. The Bloemfontein City Council, in South Africa, adopted such policies in the 1920s (La Grange 1986), as did Britain (Payne 1984). India, Indonesia and Turkey employed self-help programmes in the 1950s (idem.). However, these examples were relatively isolated and the more common response was to ignore research findings and new orientations (Ward 1982). Western academics expounding on the attributes of squatter settlements, whose inhabitants were distrusted, did not generally receive an enthusiastic audience and elite stereotypes regarding the poor were not easily overcome. In addition, public housing had become the province of the construction industry, whose interests were threatened, while the houses so provided were frequently inhabited by state employees, who could therefore be expected to resist a reorientation in housing policy.

The adoption by international aid organizations of a basic needs focus in the early 1970s, and their making resources available for self-help housing, did much to change the picture. When a conservative institution such as the World Bank offered soft loans for self-help projects, it not only lent more credibility to these projects, it also reduced the risks involved and made it profitable for the construction industry to participate in a still lower (but not the lowest) end of the housing market. A large proportion of self-help housing projects around the world have been funded by international aid institutions (Payne 1984). Between mid-1972 and mid-1982 the World Bank provided $2.5 billion in support for the creation of 310,000 site-and-service lots and the upgrading of 780,000 existing dwelling units. It is estimated that this has benefited more than 10 million people (Baum and Tolbert 1985).

AN ASSESSMENT
Self-help housing is now appraised more cautiously. From Turner's point of view, self-help housing and the earlier recommendations were not intended to lead to a series of individual projects. Rather, there was to be a general legal and market environment that would enable a myriad of individuals and institutions to become involved in the housing process. A few projects do not serve to change the fundamental imbalance in the supply of, and demand for, land.

Another criticism was that self-help projects seldom reached the poor. The cost of the projects and the inadequacies in the supply, which keep prices up, bar the poor from access to sites. As noted, when, in a few instances, the poor were allocated sites, the land was usually promptly reallocated by the market in favour of the better-off (Harris 1972). It was found that the poor are particularly quick to capitalize on such windfalls. 'Perverse' outcomes are also likely where the poor are more inclined to rent living space rather than supply their own housing, as in Lagos or the 'bustees' of Calcutta. In such instances, improvements resulting from squatter upgrading schemes increase the attractiveness of the area and result in rent increases. Higher rentals (or plot sharing) are important for the landlord because they represent a means of financing the expansion of the existing building or the start-up of another dwelling unit in a new squatter area (Payne 1984; Peattie 1987), but the poor, again, may be caused to move on.

In addition, and somewhat ironically, self-help housing has been subjected to much the same criticisms earlier levelled against public housing. These are that 'sites-and-services schemes were often poorly located in relation to the main income and employment sources of low-income groups. To reduce costs, they have often been developed on the city periphery where land was cheap and easy to obtain. Many schemes only provided one or two choices in terms of plot size, plot price and repayment conditions' (Sattherwaite 1983: 51,52).

The heyday of site-and-service projects is over. Self-help housing has benefited many, but a project-by-project approach does not allow for a sufficiently rapid increase in the supply of housing. The answer here is not simply that one must increase the number of projects – a shortage in staff and capital usually prevents this; instead one should strive to open up the housing process so that self-help can be taken more literally. Even more important is the fact that funding for self-help housing is drying up. The demise of self-help housing reflects changes in the World Bank's development orientation, where the concern for basic needs has become outmoded. As sketched out by Mabogunje,[6] the objectives of urban policy are no longer schemes addressing basic needs in cities (most prominently housing), but rather focused governmental intervention on 'city performance'. How should the city's growth be managed? With economic development in mind, what investment in infrastructure and services and residential development is optimal? Housing remains an issue, but its prioritization is to be determined using economic criteria.

Lastly, the self-help housing phenomenon, which occurred independently of and frequently prior to government attempts to harness the process, was possible because of access to relatively low-cost land. This precondition to the process is less and less available. Thus, '[A]ll over the world, markets are being consolidated into fewer and fewer owners, who will have correspondingly greater power over its disposition. In many cities an era has ended, and policy-makers can no longer depend on the flexible structure of the city, and its institutions and its landowners to provide an accessible supply of land' (Doebele 1987: 116).

Public–private partnerships

The current policy environment lacks direction, except in that a greater role is envisaged for the private sector. This is a useful orientation given World Bank research, which identifies strong regularities between GDP per capita and the proportion of the national income devoted to housing, regardless of whether the expenditure is undertaken by the public or private sectors (Mayo 1988). This suggests that if the public sector undertakes housing construction it will largely displace private sector construction. It can reasonably be objected that the private sector will serve a different market from the public sector and that the same expenditure by the public sector will produce more houses for lower-income groups, except that public sector construction has itself commonly not been allocated to the poor, in both socialist and Third World countries (Szelenyi 1983; Macoloo 1988). An increasing private sector role is also important because it brings greater management expertise and efficiency, and more capital, to the housing problem. It is therefore worth examining the nature of public–private partnerships, which are the usual manifestation of an increased role for the private sector. These partnerships, however, are invoked more often than they are defined or practised and I have therefore referred to the Second International Shelter Conference and the Venter Commission for clarity.

THE VIENNA CONFERENCE
A prominent depiction of the nature of a public–private partnership is provided by the 'Vienna Recommendations on Shelter and Urban Development' prepared during the Second International Shelter Conference held in Vienna in 1986. The conference was funded by, among others, the National Association of Realtors (USA), the International Union of Building Societies and Savings Associations and the International Real Estate Federation, so there can be little doubt regarding its private sector credentials.

The Vienna conference was called the second conference as it was a response to the Vancouver United Nations Conference on Human Settlements mentioned earlier. The 'Action Plan' drawn up at the first conference is held to have been over-reliant on public sector programmes that failed owing to limited institutional capacity and inadequate financial resources; and also owing to the assumption that the best way to approach the housing shortage was a series of projects rather than a favourable legislative and policy environment that facilitates myriad, individual housing investments. In contrast, the 'Recommended principles for action' of the Vienna conference include:

- forms of land tenure which create efficient and cheap procedures in respect of 'the land market [and] equity in the form of access to all groups requiring land . . .';
- public policies 'which provide incentives to develop urban land for various uses in response to effective demand';
- 'land use standards . . . which reduce development costs and facilitate the access to land of low income groups'; and
- 'government intervention in private land should be limited to zoning, acquisition for public purposes with just compensation and in exceptional circumstances to facilitate access to land for low income groups'.

These provisions do not assume, and are not directed at, the creation of a role solely for the formal private sector. The provisions, and the Recommendations in general, laud the role of informal forms of housing supply. In effect, what is being advised is the incorporation of these informal systems into the formal system through reducing the costs of participation in that system. Where does all this leave the urban poor?

In respect of land, Van Huyck (1987: 21), who was prominent in organizing the Vienna conference, urges that 'Governments need to facilitate the timely supply of land for urbanization at the rate and scale required through the use of incentives and taxation instruments which induce private landowners to develop urban land in response to effective demand'. The problem, of course, is that, since it seems that the poor are unable to afford land when there is a free market in land (Connolly 1982; Payne 1984), the benefits of privatization for the poor are not especially evident. Even though public housing and site-and-service efforts have a poor record, privatization seems to involve a degree of cynicism, since to make social programmes solely dependent on private sector supply is to deny the poor access to the benefits of that programme – privatization has its limits.

An additional problem concerns the point that the private sector may itself lack the resources to engage in the housing sector to the level needed to redress housing shortages. For example, Bob Tucker, managing director of the South African Permanent Building Society, has estimated that the capital required to house South Africa's black population in the year 2000 will be more than four times greater than the capital available to the country's banks and building societies. I argue below that this scarcity is inevitable in less developed and middle-income countries and one can be certain that privatization will not relieve the poor of their desperate situation.

THE VENTER COMMISSION
The problem addressed by the Venter Commission related primarily to the cost of providing serviced land (South Africa 1984a). The vehicle recommended to achieve the necessary scale of investment was a development corporation 'whereby funds from the private and public sectors are jointly channelled into the development programmes of local authorities' (5.2.1 (f)). To this end it seeks a *'formal partnership'* between the two sectors 'within the framework of a township development corporation'.

The Venter Commission proposed that the public sector should bring to the partnership the:

(a) Speeding up of administrative and other processes involved in township establishment. Allied to this is the engendering of a positive spirit of cooperation.

(b) Development capital. . . . the Commission recommends
     that the State . . . make the maximum amount of funds
     available for the provision of residential sites and hous-
     ing.
(c) Methods of curbing costs . . . , eg. [reducing] development
     standards and making [State-owned] land available
                                                     (5.3.1)

The private sector is held to offer:

(a) Development capital.
(b) Expertise in the field of management, financial control
     and operation.
(c) Entrepreneurial skills.
                                                     (5.3.2)

The goals of the development corporation would be:

(a) Operating on an economic basis
(b) The speedy execution of development projects.
(c) The steady supply of building sites, including housing
     where applicable, to the lower and middle income groups
                                                     (5.3.3)

The Commission viewed the private sector as the primary vehicle
through which the housing crisis was to be resolved. This is
revealed by the subsequent point that the 'private sector should
provide a larger amount of the share capital than the State' (5.3.4
(a)), and would 'have greater representation on the board of direc-
tors' (5.3.4 (b)). This partnership would leave the high end of the
market to the private sector and seek to negotiate a *modus vivendi*
for the low end of the market where the private sector can realize
profits and costs can be restrained. In order both to supply sites or
housing at cost to the lower-income groups and to allow the private
sector to obtain the going rate of return, the Commission proposed
that the government provide the corporation with interest-free
loans, reduce standards, supply public land free, not charge for
administrative costs, act to reduce risks, assist low-income families
(unspecified) (5.3.7). 'Certain disciplines' would also be applied in
respect of middle-income families and some of the above measures
would be repeated.

One can dispute many aspects of the Venter Commission's
proposals; for example, the attempt to pass the capital costs on
to the private sector promises no relief for the really poor. To
hold that the private sector can profit from providing the urban
poor with serviced land *and* that the land will be affordable is

make-believe. The public sector would have to bear considerably greater fiscal pain than envisaged by the proposals. Yet there are aspects to the Commission's proposals that, if implemented within a democratic environment, would be of some interest. In particular, the proposals intended to reduce the cost of developing land and to finance urban development are worth a second look. I return to these issues in the conclusion of the chapter.

Democratic preconditions

This section focuses on, not whether governments can address the housing needs of the poor, but what prompts them to seek to do so. Middle-income countries cannot provide sufficient formal housing because they combine:

- severe demands on the fiscal resources of the government;
- limited economic capacity, for example, a lack of artisans, equipment and management expertise;
- low incomes such that there are many who cannot afford a house or serviced site, a problem that is exacerbated by relatively high levels of income inequality; and
- relatively rapid population growth and increases in the urbanization level.

This view is reinforced by the close relationship between GDP per capita and the share of housing in GDP. Ordinarily the share of housing in GDP varies between extremes of 1 and 7 per cent, with the higher measure being characteristic of middle-income countries with growing economies (Renaud 1987). South Korea illustrates the point, in that it was only after many years of rapid growth that, in 1981, the fifth Economic Development Plan was renamed the Economic and Social Development Plan and the targeted investment in housing increased to up to 6 per cent of GNP. By way of comparison, the contribution of investment in housing to South Africa's GDP was 3.2 per cent in 1960, in 1970 it was 4.2 per cent, in 1980 it declined to 2.5 per cent and in 1988 it was 2.55 per cent. What conditions must prevail before the South African government actually delivers housing to the poor?

Are housing problems in the First World intrinsic to capitalism and, similarly, is the extent of the shortage in middle-income countries due to a capitalist mode of production? For example, is it correct that the 'housing problem [is a] structural condition of the capitalist mode of production' (Burgess 1982: 77); or that 'As long as the capitalist mode of production continues to exist it is folly to hope for an isolated settlement of the housing question affecting the lot of the workers' (Engels 1970: 74). Is it true, therefore, that 'The solution lies in the abolition of the capitalist mode of production and the appropriation of all the means of subsistence and the instruments of labour by the working class itself'(idem.)? The following examination of housing policy in First and Third World countries, both capitalist and socialist, contradicts this position. While housing problems assume certain forms under capitalism, they are neither exclusive to capitalism nor necessarily even more pronounced under capitalism.

If neither capitalist nor socialist countries do a good job of housing the poor, are there other conditions under which governments do, in fact, fulfil this function? In this section I set out to elicit the conditions under which governments serve the poor.

There are few countries that do not have housing problems. This chapter was written in the USA and I had only to open the *Boston Globe* to witness the ongoing struggle of the poor to obtain decent housing. Yet Western European examples make it clear that adequate housing is possible in developed, mixed economies wherein the interests of the poor have prevailed. It is surely self-evident to assert too that, in countries such as the USA, inadequate housing reflects the outcome of political–economic forces and that, in this light, it is no surprise that under Reagan homelessness increased so dramatically.

The same is not true for Third World countries. Here political–economic struggles affect the distributional parameters of the problem but, owing to the inability of the means of production to deliver sufficient houses, not the existence of the problem per se.

Is the situation any different in socialist countries? Certainly one cannot hold that here housing problems are banished. Cuba, for example, forbade the supply of formal private sector housing (in 1963) and emphasized government-supplied housing (Gilbert 1982a). The Cuban government's intentions notwithstanding, it was unable to meet demand. The 1970 census showed that there were 600,000 more houses than in 1953, of which only 200,000 had been built by the government (Peattie 1987). However, the absence of a housing problem is too demanding a criterion by which to assess socialist countries that themselves have limited economic capacity. The more appropriate question is whether the distributional incidence and the extent of the problem are reduced?

To the extent that it is possible to generalize about the housing policy of socialist countries, one can say that, in practice, social-ist states have themselves accorded a low priority to housing expenditure and have allocated the housing that is constructed in an inequitable manner. Thus, Eastern European countries have viewed housing 'as unprofitable expenditure – as consumption of national income, rather than as productive investment' (Szelenyi 1983: 24). The consequence of Eastern bloc countries favouring investment in industry is that a considerable proportion of the housing construction that does occur, occurs privately. What is more, the private housing construction is undertaken largely by the poor, because those with better incomes and positions in the state apparatus are invariably first in line for new state housing and those employed in state enterprises are frequently provided with housing by those enterprises. Administrative allocation of housing serves to reinforce, rather than counter, existing social differentiation. Since the governments have, in addition, subsidized the cost of this maldistributed housing, the result has been fewer resources available for further investment in housing. Not only have housing subsidies not been available to the poor, the use of subsidies has reduced the capital available for additional construction.

This Eastern European example, where the relatively well-off are housed by the state and large enterprises and the poor have to house themselves, parallels conditions in capitalist Kenya. In Kenya, 'a much higher proportion of people in the lowest income brackets are forced to find housing in the private sector. City council, government and private firms make housing available to persons in middle and upper income brackets' (Harris 1972: 44). On the other hand, both examples are contrary to housing practice in Western Europe, where the rich house themselves and those subsidies that are made available, are directed to the poor. These examples, certainly, give one reason to pause before generalizing about the differences between capitalist and socialist countries.

Within a democratic environment, there should in fact be no prior presumption that either Marxist or reformist governments have any intrinsic reason to serve the poor. Marxist parties are inclined to treat the poor in a jaundiced manner: as a generalization, the urban poor are not made up of labour from the formal sector and their isolated occupations, in small enterprises, do not predispose them to organization as members of a class. In addition, it would be incorrect to equate the interests of, for example, all those living in shantytowns (Nelson 1979). The poor are themselves stratified. Thus, in Chile, the 'Left argued that the *poblador* lacked experience of proletarian organization and therefore would have a low level of political consciousness, and furthermore that organizing around demands for consumption would lead to economist deviations, since *pobladores*, who had no experience of a struggle against the bosses, would fail to visualize the bourgeoisie as a class' (Threlfall 1976: 170).

The urban poor have also attracted little interest from reformist parties throughout the Third World. Nelson (1979) explains this as being due to the following reasons. Historically, the urban poor have constituted only a small proportion of a nation's population. This, obviously, is nowadays characteristic only of African countries. Secondly, elitee attitudes 'are dominated by disdain, fear and ignorance about the poor and their way of life, sometimes leavened with paternalistic concern' (p. 320). Such 'stereotypes about migrants and the poor provide convenient rationalizations for avoiding fundamental reforms and radically altered priorities' (p. 321). Thirdly, a party that attempts to appeal to the poor will usually have to develop a platform that conflicts with the interests of other, more influential, constituents. Lastly, the nature of political competition may facilitate other means of mobilizing sections of the poor. This is especially the case where the political competition is largely based on ethnic competition: where ethnicity cannot be utilized to mobilize support, the poor are likely to fragment their support. In Chile, for example, Allende's support came more from organized labour than from the workers with more precarious income sources, who leaned more to the right (Briones, cited in Nelson 1979). There can be no presumption that the poor will autttomatically support a socialist party.

It is instructive, though, to consider the targeting of housing benefits in Chile under Allende. 'In general, the housing programme of the Unidad Popular [Allende's party] was structured so as to benefit, in the first place, the lowest income groups living in the large squatter settlements around every city, and in the second place, the regularly employed working class . . . 80 per cent of the new units planned for 1971 were allocated to the lowest income groups' (Lozano 1975: 186, 187). This targeting did not correspond to the location of Allende's support, which was principally among organized industrial workers, and arose as a result of the fact not so much that Allende won the 1970 election (receiving only 36.2 per cent of the vote), but rather that splits within Frei's Christian Democrats and other disputes lost it (Nelson 1979: 351). It was therefore critical for Allende to gain greater support among the poor if he was to win the next election.

With respect to differences in the programmes of the Frei and Allende governments, the former 'championed the cause of the "marginals", the rural and urban poor and *lumpenproletariat* and those who inhabited Chile's myriad squatter settlements or *poblaciones*, as an alternative to the organized working class, claiming, with some justice, that the traditional parties of the Left ignored these sectors'

(Roddick 1976: 4, 5; emphasis in original). The urban poor had supported the Christian Democrats in large numbers; thus, despite the fact that Allende's own convictions and those of the Communist Party emphasized the need to gain the support of the 'middle sectors' (Roddick 1976: 15), it was an electoral imperative that the support of the poor had to be weaned to Unidad Popular.

It seems that three conditions must be present before parties seek the support of the urban poor and before governments deliver to the poor. The conditions are:

• a large urban population relative to the national total;
• stiff electoral competition; and
• an ideological predisposition on the part of one or more of the competing parties to the participation of the poor.

It was just such conditions that obtained in Chile. It is such conditions that one hopes for in South Africa.

In summary, one should not expect that housing programmes will, or will not, benefit the urban poor in both capitalist and socialist systems. A different condition seems critical if housing policy is to benefit the poor, namely that they need to constitute a significant constituency. Thus, 'The spontaneous settlers of Lima, Anakara and Caracas have all been the beneficiaries of political competition' (Gilbert 1982: 111). In the hope that one can assume such a situation in South Africa, to hold that public schemes to alleviate the housing shortage represent 'proposals for the maintenance of the capitalist mode of production' (Burgess 1982: 86) is to trifle with the situation of the poor. It is reasonable to pursue housing programmes in the belief that they 'can effectively bring about some measure of improvement in living conditions, even though they cannot alter the more fundamental causes of urban poverty' (United Nations Centre for Human Settlements (HABITAT) 1981: iii).

Policies intended to increase the supply of land

The rest of this chapter is devoted to the discussion of policies that are intended to increase the supply of land for the urban poor. My presumption, like that of Blumenfeld (1986: 28) following his examination of the resources available to the private sector, is that 'private sector involvement in black housing development is unlikely to come to very much'. Since the very poor are unable to afford the cheapest house the private sector appears able to provide, and most can also not afford a serviced site, I am here concerned with how best the government can intervene. We can be certain that a democratic government will be under intense pressure to increase the supply of land and housing and that the traditional town planning regulations that affect density specifications will prove too sedentary. We can assume that the government will adjust these regulations to its ends and then seek additional measures.

The problems I address focus on how to increase the supply of land, how to restrain price increases and how to promote increasing urban density and infill; the policies I debate involve indirect interventions in the land market and direct interventions in the form of a LandCorp. In large part the first alternative involves the use of various forms of taxes. The second alternative includes reference to land readjustment programmes and land banking.

The market for land on the Witwatersrand

This study initially had, as one goal, a determination of the potential supply of land, but this objective has long since been abandoned as inordinately static. Supply does not exist in an abstract sense; it is directly related to the price of land and to a multitude of other considerations. There are far too many uncertainties to put a figure on the amount of land available. Yet, if one disregards racial zoning, I am unaware of any observer who has identified constraints to the physical supply of land as a problem for housing. For instance, the Department of Transport (South Africa 1987: ii) comments: 'Due to the existing relatively low development densities, well-developed mass transportation systems (for low-income people) and a surprisingly ample supply of developable land, there is reason for considerable confidence in the ability of our cities to accommodate the expected growth in population . . .'. This is in contrast to many other Third World cities in respect of the supply of land:

> The majority of people who came to large cities in developing countries in the last two or three decades found or developed housing in popular settlements. It was a historic epoch of non-commercialised or cheap commercialised land supply . . . The evolutionary process of low-income housing development was made possible and conditioned by the supply of land for which people did not have to pay or paid very little. This has been a temporary phenomenon in modern urbanisation . . . this era in many countries is drawing to an end.
> (Baross 1983: 205).

But, while the PWV is not short of land, the number of actors in the land market are rather limited and this creates considerable potential for collusion and the withholding of land in order to bid prices up. The market interventions described later are intended to cause owners of this land to bring it onto the market and to use it more intensively.

The key features of the market are shown in Figure 4.1: the structure of the Witwatersrand, the low density of the urban areas, the extensive mining land that is held vacant, and the (large and small-holder) farming land that is close to the city and has been

Figure 4.1   The density of, and land availability on, the Witwatersrand

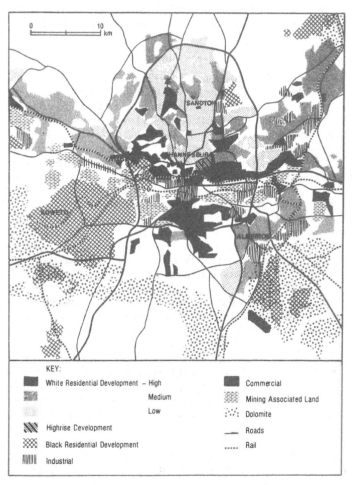

KEY:

White Residential Development – High	Commercial
Medium	Mining Associated Land
Low	Dolomite
Highrise Development	Roads
Black Residential Development	Rail
Industrial	

left unshaded the figure.[7] In respect of the white urban areas shown in Figure 4.1, Boden (1987: 7,8) has argued that there is an 'innermost zone of established suburbs', many of which are undergoing intensification pressures; and further out there are 'schizoid' outlying suburbs that display a mixture of medium-density cluster or group housing and spacious neighbourhoods with managerial villas. Market interventions should be directed at promoting the increasing subdivision of properties and buildings and a decrease in floor space per person, as space-extensive houses and apartments become too expensive for individuals and couples to afford. With the demise of residential segregation, blacks, coloureds and Asians, who generally have larger families, would (it is hypothesized) occupy hitherto 'white' residential units at higher densities. This would especially be the case when families share previously single-dwelling units.

The likely future changes in the established black townships are less clear. For example, while there is tremendous overcrowding in the dwelling units, the townships have a low built density. Does this mean that they can sustain 'infill'? It appears that there might well be resistance to this process, many of the seemingly vacant spaces having gained the status of 'peoples' parks'. None the less, density increases are likely as a result of renting rooms or sharing houses and backyard shanties. (Many formerly white areas might also acquire these shanties.) Yet, while I would argue that greater density is desirable in general, this is not necessarily the case in the townships. This is especially evident in the sharing of residential space, since it is already held that, in Soweto, up to 16 persons share sites that contain a house. (When this occurs it is likely that some live in associated structures.) The basic house (model 51/6) contains four rooms including the kitchen, and has a total of 40.4 m² internal space. Indeed, one of the means employed under apartheid to restrict access to the cities has been overcrowding to the point of intolerability; to the point, in other words, where people prefer to live further out and spend a larger proportion of their low income on commuting, rather than on food, education or shelter.[8]

High densities are also likely to emerge in squatter settlements on the fringe of the city. The LandCorp proposal that follows is intended to enable the government to create opportunities for such settlement through the provision of serviced land.

The second main feature of the market concerns the land held by mining companies. Figure 4.1 shows a great deal of mining land extending along an East–West axis through Johannesburg. Apparently 'More than half this land is unused and could therefore possibly be used for non-mining purposes' (South Africa 1986b:

180). Not all the land would be available for housing since some is undermined or underlain by dolomitic geological conditions and other land would have an optimal use and a market value that would lead to other land uses. Nevertheless, there is equally much land that could be beneficially employed for housing the poor. At present, the mining companies are not taxed on the land so long as the land is held for ostensible mining purposes, leaving them in a position of bearing no cash cost while, at the same time, being able to speculate against future increases in the value of the land.

The remaining land indicated on the figure is used for farming or smallholdings, which have an average size of about 2 hectares, or may not be employed at all. Except for some valuable market gardening around Germiston, the land is in general poorly endowed for agricultural purposes and its value largely arises from its potential for inclusion in the urban centres. Here, too, is a large potential supply of residential land.

A last point follows from the fact that vacant mining land and land zoned for agriculture have not suffered from the increase in price that follows on a change in zoning and use. There is considerable potential here, a 'window of opportunity' almost, for a government to intervene in the land market so as to promote an increase in the supply of reasonably priced land (La Grange 1986).

Indirect intervention: the reason for taxing land[9]

The use of taxation incentives or disincentives is typical of a hands-off policy where the government attempts to influence, rather than direct, market outcomes. Taxation of land is primarily intended to generate municipal revenue, but, while this is certainly an important concern, it is not the issue being addressed here. Rather, the focus is on the use of taxes to induce certain land uses, to bring forward in time decisions to sell land, to reduce the cost of land and to inhibit speculation. Tax policies embodying these goals are common throughout the world and the taxes examined below are: site value taxes, vacant land taxes, capital gains and transfer taxes, and betterment levies. Since the influence of these taxes is determined by expectations regarding the rate of appreciation of land values in comparison with alternative investments, their purpose is to create a predisposition to certain decisions with respect to land.

The goals against which the use of these taxes should be measured are prompting holders of land that is currently un- or underutilized either to sell or to develop that land, and restraining or lowering prices and inhibiting speculation. An additional goal, frequently associated with taxes on land, concerns taxes on 'unearned increments in land value'. To clarify, urbanization is a process that creates wealth and the process is especially manifest at 'stages of transition' – when rural land on the urban fringe is put into urban use, when public investment in infrastructure and services attracts higher-value land uses, and when there is change in permitted land uses. In each case, the wealth created is for landowners, who benefit without any entrepreneurial activity on their part. Taxes on unearned increments of this sort may either be a capital gains tax on transactions on land or the much more limited betterment levy, intended to recover for the public sector the cost of its investment in infrastructure.

SITE VALUE TAXES

The tax system of municipalities in the PWV employs the site value system where the valuation is based on the market value of the land, irrespective of what is on it. A valuation is reached by examining recent market transactions, either of the particular site itself or of similarly located sites with equivalent qualities. The tax in the PWV does not include services and additional amounts are assessed for water and electricity consumption, refuse removal and the number of sewerage connections. Within each municipality, the tax rate (a proportion of the value of the land) stays the same regardless of the value of the land.

While the disadvantage of the site value system is that it usually does not generate as much revenue as the other municipal tax systems, it does embody distinct advantages. One is that it does not act as a disincentive to development or to the maintenance or improvement of buildings – on the contrary, it serves to promote development on the land.[10] Another is that the value of the land can be determined relatively objectively, being based solely on market trends.

Site value taxes reduce the market value of the land because the tax increase is capitalized.[11] This means that a prospective buyer of land will lower the price offered for land in relation to the effect of the tax on the present value of the land, as determined by the discounted future earnings anticipated from the land. For example, if the current discount rate is 10 per cent, a R1 increase in taxes in perpetuity may result in up to a R10 decrease in the price of the land. The effect of site value taxes on the timing of decisions to sell the land is that land will be held for as long as its rate of appreciation plus its rate of return from current use, minus the property tax rate, is greater than or equal to the rate of return of alternative investments.

An increase in site value taxes has two effects. One is to cause a drop (due to tax capitalization) in the amount developers will be willing to pay for the land in the future. The second effect is to cause the opportunity cost of holding the land to change, either upwards or downwards, which means that the effect of changes in site value taxes on the timing of decisions to sell the land is therefore uncertain. It may seem paradoxical that an increase in taxes produces this uncertainty, but this may be explained as follows. The tax causes the capital value of the land to fall, but an unchanged income stream means that the return on the capital value rises. Depending on the relative values of the site value tax and the income generated from current use, the opportunity cost of holding the land may rise or fall.

If the immediate effect of the tax change is to make the holding of land for speculative purposes unattractive, then the increased supply of land made available to the market may restrain price increases. The longer-term effect, however, is that the site value tax having been capitalized, the rate of price appreciation may remain unchanged and there will be equally little incentive for a decline in speculative activity.

An alternative interesting application of site value taxes concerns their being structured in some countries in such a way that, unlike in the PWV, they increase with property values. The redistributional consequence and cross-subsidy effect of this is obvious if, say, persons living on properties with an extremely low value are not taxed at all and there is a progressive increase in rates as one moves up a scale of property values. Taiwan, for example, employs a tax that ranges from 1.5 to 7 per cent of the value of that land. This tax makes sense when applied to residential areas, in that it promotes infill by predisposing householders to live on smaller pieces of land. Its value is less obvious when applied to, say, industrial or commercial areas where the problem may be more one of how to promote the consolidation rather than the fragmentation of land holdings.

VACANT LAND TAXES

The primary purpose of site value taxes is to raise revenue for municipal governments and it is seldom the case, as in Taiwan, that their main function is to 'induce particular land use effects' (Harris 1979: 191). While these taxes create a desirable predisposition to the development of land, it must be recognized that they are a blunt instrument for this purpose. If one's objective is to promote the sale, at reduced prices, of particular types of land and then the subsequent development of that land, an increased site value tax penalizes all existing and future landowners throughout the city.

Instead, it may well be preferable to employ vacant land taxes on top of a site value tax. The tax does not require that the land is literally vacant – from the point of view of a desired intensity of development, it is sufficient if the land is simply underdeveloped.

Vacant land taxes are widely applied, being employed, for example, in Singapore, Lusaka, Nairobi, Jakarta, Bombay, Abidjan and Calcutta. The Singapore example is illustrative of their application. Here, specified vacated plots and properties that have an unusually low building-to-land coverage (Bahl 1979) are assessed at 5 per cent of their capital value, which is more than twice the rate applied for improved properties.

The effects of a vacant land tax have been described by Smith (1979: 151) as follows:

A vacant land tax is unlikely to be fully capitalized, since 'vacant' urban land is generally in a temporary state – the land will be developed in time. Since the vacant land is worth less to those wishing to hold it idle, or semi-idle, the developers may now be able to make a purchase at a price which the owners previously rejected. Thus *land becomes available for development at a lower price* . . . *[and] at an earlier point in time.* (Emphasis added)

The timing effect emphasized above is important for the following reason. As the vacant land tax ceases to apply once the land is developed, the potential development value of the land does not change. The rate of return from the current use will also not be affected by the tax. However, the cost of holding the land rises, since there is now an additional tax on the land and prices will be expected to fall.

Vacant land taxes are a more selective instrument and do not have the extremely unpopular consequences that an increase in site value taxes involves, namely a decrease in property prices for all – it is much more discriminating with respect to victims and beneficiaries. Developers, the construction industry, and future home-owners benefit from the tax and only certain predetermined original landowners are left worse off.

CAPITAL GAINS AND TRANSFER TAXES

The above, of course, dwells in the realms of economic theory. The 'many other influences on the redevelopment decision complicate the measurement of the pure tax effect' and the taxes just described are one among a number of factors that need to be considered when making decisions about land (Bahl 1979: 41). It may seem appropriate, therefore, to approach the problem more directly. Does a capital gains tax on increments in land value not serve to lower land prices and to inhibit speculation? Would it not capture for the public the benefits, formerly accruing to landowners, of increases in value resulting from public investment, and increases in demand and rising prices resulting from population and economic growth?

There is a mixed response to these questions. A capital gains tax reduces the after-tax price one can expect when land is sold and, since potential buyers realize that they too will be taxed on future price appreciation, they in turn have reason to offer less for the land. The value of the land is reduced for both existing and future landowners and the effect *should be* to lower prices.

One reason this conclusion may not apply depends on whether the tax is applied as it accrues (when the tax is imposed) or when it is realized (when the property is sold). If the tax applies in the latter form and can be put off by not selling the property, then the withholding of property from the market may cause prices to be higher than they would otherwise be. This outcome is referred to as the 'lock-in effect' and is contrary to that intended. But this is not to say that the alternative of taxing land when the benefit accrues is preferable, since it imposes a burden that many will be unable to meet – they may be forced to sell their property. Moreover, Smith (1979) has described the administrative problems associated with this form of tax as 'insurmountable'.

Another limitation to a capital gains tax is that there is little reason, once the tax has taken effect, for price increases to slow down. Just as is the case with a site value and to some extent with a vacant land tax, tax changes are capitalized and reflected in the value of the land. Thereafter, changes in the value of the land are driven by demand resulting from the factors already noted – urban growth, public investment in infrastructure, and the conversion of land from one use to another. The attractiveness of investing in land depends on price changes relative to the return available on other possible investments. From the point of view of the speculator, there is not now less reason to consider investing in land.

Finally, there are also transfer taxes that are specifically directed at inhibiting land speculation, such as have been used in Taiwan, Korea and Kenya. Transfer taxes are expressed as a percentage of the sale price of the land. They are expected to make transactions in land less attractive because, for a transaction to occur, the buyer must believe the land to be worth more than does the seller, by an amount at least as great as the size of the tax. If the tax is large there will be fewer instances where the views of buyers and sellers are at such variance. If, however, the market is relatively large, with many buyers and sellers and prices appreciating relatively rapidly, the tax is likely to have minimal effect.

Smith (1979: 158–9) is derogatory regarding both transfer and capital gains taxes that apply only to land, as opposed to a more general capital gains tax. He writes:

> Transfer taxes disrupt transactions which would lead to the more productive use of land, usually raise little revenue, and may in the short run result in higher prices and postponed development for vacant land if the lock-in effect is significant. Taxes on increments in land value have proved to be ineffective on a number of occasions. In the case of the most recent British experience with this type of tax it was far from clear whether it led to an increase or decrease in land prices. . .

BETTERMENT LEVIES

Betterment levies are taxes on unearned increases in land values that result from proximate investments in urban infrastructure. Their purpose is not to tax the full extent of the increase, but rather to recover the costs of the public sector's expenditure, which is essential if the public sector is to be able to undertake further investment in infrastructure. The need for betterment levies is due to the fact that, since site value taxes are low and lag behind market prices, they typically capture only a small proportion of the increase in value that arises from public investment in infrastructure.

A high rate of investment in infrastructure, in the creation of serviced land, is imperative since the primary cause of high land prices is an inadequate supply of serviced land (Doebele 1983: 79). The problem has two parts. One the one hand, an increase in the urban population and economic growth lead to a rapidly growing demand for land for houses, social facilities, shops and factories. On the other hand, efforts to increase the supply of land with water, sewerage, electricity and roads are costly and slow. Governments invariably fail to increase the supply at the rate necessary to cope with migration and population increases. The imbalance between the supply of, and demand for, serviced land leads to rapid price increases.

The surplus value of land is the increase in the capital value of land that exceeds an allowance for inflation, holding costs, entrepreneurship and risk. In South Africa there is already a levy on increases in value that result from changes in permitted land use. For our purposes – namely, how to increase the supply of serviced land – the pertinent issue is how the government recovers the cost of providing services; the goal is to enable it to provide serviced land at a higher rate than would otherwise be the case.

Betterment levies intended to recapture project costs have been successfully applied in some instances. For example, in Bogota, Colombia, 'Street improvements, sewer extensions, and other services have been financed . . . by valorization charges, a system of taxation by which the cost of public works is allocated to affected properties in proportion to the benefits conferred' (Doebele *et al.* 1979: 73). However, as might be expected where money and taxes are concerned, there is complexity. Shoup (1983: 143) comments that 'Financing public works by betterment levies is equivalent to requiring owner-occupants to receive a public service and pay for it (through the levy) or else move out, realizing whatever net capital gain remains after payment of the levy'. Having to move out is frequently the only option available to the poorer inhabitants of an informal settlement – they realize the benefit of improved services but are landed with a cash flow burden that they may not be able to meet. The same applies when they are renting property and the improved services enable the landowner to charge higher rents.

There is no obvious means of countering the problem of higher rents for the poor. One way of attempting to cope with the impact of the betterment levy when the poor own the land is to impose the levy when the benefit is realized, that is, when the property is sold and not when it accrues to the landowner. In this case the levy would be equal to a tax imposed on the sale of land and the obvious outcome would be the 'lock-in effect': land would be withheld from the market and, owing to a shortage of land, the rate of price increase might accelerate.

Shoup (1983: 145) has suggested what is probably an administratively complex but otherwise attractive means of implementing the levy without the above negative effects. His suggestion consists of allowing property-owners to defer the payment of special assessments, with accumulated interest, for so long as they own the properties. This would enable the owners to benefit from services without confronting cash flow problems and, since their property values will come to reflect the services available, they will profit when they sell. From the point of view of local government, the government is, in effect, offering the property-owner a loan to pay the assessment. At the same time the government does not lose since, if the owners are charged a market rate of interest, the present discounted value of future payments will equal the initial special assessment.

Betterment levies have worked in Colombia but, even there, cost recovery has been more successful in the case of simple projects like a road or a park, where the benefits are directly felt by adjoining owners, and where the owners have participated in the planning of the project (Doebele *et al.* 1979). It seems that betterment levies are more appropriate for some projects rather than for others and that, whatever the project, it is seldom the case that one can actually hope to recover all of the costs entailed (Grimes 1977). Elsewhere, the application of betterment levies has stalled owing to the complexities involved – this has especially been the case when disputes arose over the benefits realized by the landowner.

An alternative to betterment levies is provided by service charges, but only when the cost of providing a service can be met in the form of a user charge or tariff. 'Lump sum charges' for a service are not included here since this is equivalent to a betterment levy except that there is usually no commitment to proportioning the charge in terms of the benefit received. Lump sum charges face the same problems as betterment levies and, unless the charges are in proportion to the benefit, are inequitable and have a regressive impact. Since the poor are usually unable to meet lump sum charges, the capital costs are frequently included as part of a single tariff of user charges (Dunkerly 1983: 24), but this method does not escape the criticism that it raises the cost of living for the urban poor and may cause their displacement from the area.

A South African response to these problems is illustrated by the electrification programme for Soweto. Residents pay an initial R700 to be connected to the system, after which they pay R12 on a monthly basis to meet the R268.5 million capital costs of the programme, a R4 service levy and then also the going rate for their consumption of electricity. The government maintains that the payment necessary in order to meet the capital cost is R30, and that it is subsidizing the R18 difference; but in fact it originally intended to increase the R12 charge to R30 and dropped it in the face of boycotts.

CONCLUSION

There can be little doubt that governments need to be able to engage in cost recovery if they are to undertake an adequate level of investment in infrastructure – the question is how to go about it. It is clear that taxes have unintended effects and that they are unreliable. This acts as a warning against adopting measures with which society is unfamiliar. One should beware of the temptation, in a post-apartheid study, of assuming a 'clean slate' – the future will build on existing institutions and practices. The argument points to the continued use, in perhaps a changed form, of site value taxes, which are a part of the PWV tradition and which have a well-trained bureaucracy capable of administering them.

Yet to conclude that site value taxes should continue to be used should not blind us to the fact that taxes will have only a marginal effect on land use decisions. The intuitive plausibility of taxes has to be contrasted with the experience of their application: while they may be employed to create certain predispositions in the market, their impact is minor in comparison to expectations regarding changes in the interest rate and the returns on alternative investments.

Nevertheless, to say that taxes have only a minor impact is not to conclude that they cannot be gainfully employed and the above discussion points to a number of recommendations. One is that the progressive character of site value taxes should be enhanced and another is that there should be a vacant land tax on residential land. These two taxes will serve to raise additional revenue, will predispose private sector development to use land more intensively and will help slowly to reverse the spatially disjointed structure of the apartheid city.

An obstacle to the successful application of vacant land taxes is that much of the land likely to be targeted is currently zoned for mining or agricultural use. Mining land, however, is not taxed and the assessed value of agricultural land is so low that a tax, calculated on the basis of that valuation, would be inconsequential relative to the increases in value that would result from speculative returns. As the goal being pursued presumes changes in land use, from agricultural to residential use, or from single-residential to high-density residential, it would therefore seem appropriate to value the land as if in the preferred use and to levy the tax accordingly. The recipients of the taxes would be expected to welcome them, since the taxes will be accompanied by zoning changes and an (unavoidable) increase in land values. This increase in values may, in fact, be an important political consideration with respect to gaining acceptance of the new legislation.

Of course, a precondition for this scheme involves local author- ities being able to determine the value of the land in its alternative use. The impasse one is confronted with is an increased tax rate to be applied to land that has an indeterminate value owing to *ad hoc* city-wide changes in zoning. Given time, market transactions will reveal the effect of the changes but, in the meantime, the taxes will have to be levied. Considerable opposition and legal action would be the likely response to arbitrary valuations of the land. There is a ready solution to this problem, modelled on Taiwan's ingenious method of undertaking land valuations (Harris 1979). The owner of the property is allowed to determine the value of the land for tax purposes but this value, set by the owner, may equally be the price paid should the public sector choose to acquire the land. The latter 'threat' will have credibility only if there is a genuine possibility that the land might be acquired, and the LandCorp proposal outlined below represents just such a threat.

Direct intervention: a land development corporation

Urban development corporations have been around for a long time. They were employed, for instance, in Britain to build the New Towns and they are common in ex-British colonies in the Far East and Africa. Development corporations are also widespread in South Africa – every homeland has its development corporation and there are other corporations as well. The Venter Commission actually proposed a township development corporation in order to deal with the land shortage. However, prominent academics who commented on this text advised against such a corporation and their point of view is not to be taken lightly. Given the obvious attraction of a LandCorp – that it enables the government to intervene directly and massively in the land market – one has to pay close attention to their doubts. What are the problems associated with development corporations? In the following examination of alternative forms of direct public intervention it will be seen that there are, indeed, many negative and unanticipated consequences associated with LandCorps. Yet LandCorps offer an invaluable means of interfering in the land market and I have attempted to devise a LandCorp that avoids the problems that will shortly be described.

PUBLIC ACQUISITION OF LAND

There are a number of goals associated with the public acquisition of land. These include:

- making land available for public housing;
- stockpiling public land, which may be selectively released with a view to restraining price increases;
- using public ownership of land as a means of determining the location and form of future development; and,
- through the sale or lease of public land, recovering the costs of installing infrastructure.

While there are many advantages associated with public ownership of land, such ownership does not in itself imply a ready solution to the housing crisis. The 'experience of countries that have nationalised all land demonstrates the essential weakness of public land acquisitions – it is not the acquisition per se that accomplishes useful purposes, but rather, the policies and procedures for land use that follow acquisition' (Kitay 1985: 35). Zambia, for example, nationalized urban land – freehold titles were converted to leasehold and vacant land sites were acquired by the government (Tipple 1976) – but the slow allocation of land, continuing shortages of housing, and the high level of demand reflected in the extent of illegal squatting mean that access to land still costs a great deal. Clearly public ownership of land is only one aspect to a housing programme. When, in the absence of signals from the market or a competent administration, it leads to the slow allocation of land and its designation for sub-optimal uses, it is also an undesirable one. It must be remembered that public ownership does not in itself represent a means of redistributing wealth. 'Most forms of expropriation . . . directly benefit landowners' (ibid.: 223). Unless public acquisition of land is associated with a commitment to providing serviced land for the poor, little purpose is served by public intervention or control of the land market. I am not attempting to decry public ownership per se, but seeking out cases where it contributes to the relief of housing crises. Two prominent cases – land readjustment and land banking – are considered below.

LAND READJUSTMENT
Land readjustment techniques have been practised successfully and extensively in South Korea and Japan and have also been used less widely in other countries such as Germany and Australia. In South Korea the technique is employed by the public sector while in Japan it was employed by the private sector (until values on suburban land rose so rapidly that landowners enjoyed substantial profits without bothering to undertake this complex method of installing infrastructure). No public or private orientation should therefore be read into the possible application of the technique. I am employing the South Korean case because it is the most accessible, having been described in some detail by Doebele (1979).

A simplified description of land readjustment is that it consists of a public authority employing 'eminent domain' (the forcible acquisition of land) to gain temporary ownership of (invariably) fringe land that it decides should be converted from rural to urban use and then preparing a site plan for the area. There is, in effect, public control of the timing, location and form of development. In accordance with the siite plan, the land is replotted for private building sites and public uses such as schools and parks and the authority then installs the necessary urban infrastructure. The government retains both the land designated for public use and as many of the newly created building plots as are necessary to pay for the cost of the planning and the infrastructure. These plots are then auctioned. The original owners are allocated the remaining sites in proportion to their initial contribution, with every attempt being made to ensure that these are the original land holdings. The outcome is that the value of the land they regain will have been doubly enhanced, once by a permitted change in use and again by the provision of services.

In the South Korean case, Doebele (1979) depicts a typical scenario as being one where the public sector retains 24 per cent of the land for public purposes and a further 11 per cent to cover the cost of site preparation and investment in infrastructure. The original landowner, while regaining only 50 per cent of his land, may well still support the process. Doebele reports that South Korean increments in land value following land readjustment ranged from 300 per cent to 2,400 per cent, which more than offsets the loss of half of one's land holdings.

As one can see, land readjustment recovers for the public authority the cost of installing the services, but it is not equivalent to a capital gains tax. The value of the land returned to the original landowners is considerably enhanced but there is no intention to tax a specific proportion of the increment in land value. Indeed, the greater the increase in land value, the less the proportion of the land that will be retained by the public authority to pay for the services. In addition, the level of services implied by the technique does not suggest that plots will be affordable by the urban poor. It may be that, for land readjustment to be useful for the purpose of housing the poor, a third category of serviced land should be withheld by the public sector.

The comparable situation in South Africa is that township developers also set land aside for public purposes. For instance, private white township development practice in the PWV currently consists of the developer drawing up a plan that includes public sector (excluding central government) land requirements such as roads, parks and schools. (The developer's obligations in the last respect are being disputed.) Endowment contributions of this land, or cash, are a precondition to public approval of the scheme and the actual land designated for public purposes, or the amount of cash, is negotiated with the relevant local authority. In addition, the developer has to establish and pay for internal services such as roads, electricity and sewerage. There are other obligations for bulk services external to the township, but negotiations regarding them differ from case to case. For example, if the township will utilize 10 per cent of a reservoir's capacity, then the developer might be required to bear 10 per cent of the local authority's outstanding obligations for this item.

This township development practice differs from land readjustment because it is not undertaken for a number of small landowners

– a single developer instead will have earlier acquired ownership of all the land. In other words, land readjustment seems especially relevant when there are a large number of landowners, many of whom are reluctant to sell their land or who want to participate in the development of it. This is hardly the case on the Witwatersrand where the owners of land on the urban fringe are predominantly some agricultural smallholders and a few mining companies, so it does not seem that land readjustment is especially relevant on the PWV.

This conclusion, however, is not to deny the possibility that the public sector could bargain for land for the poor – it has significant powers. For example, the public sector could negotiate higher-value zoning designations in return for public land, but, of course, the critical aspect to the negotiations would be the anticipated increase in land values that would follow a higher-value zoning. The South Korean example reflects a situation of rapid economic and urban population growth, but such stunning increases in value are not occurring in South Africa. Developers on the Witwatersrand who obtain agricultural land, have it proclaimed as a township, install services and build small houses on plots of 250–300 square metres, can expect a net profit on investment between 5 per cent and 20 per cent. In doing so they have to cede about 30 per cent of the land for public purposes – roads, parks and such-like – and can only 'onsell' the balance of the land. (This estimate excludes land for schools.) Further, whereas the land may cost only R3–5 per square metre, the cost of installing services is likely to be R20–30 per square metre; and this does not include many other costs, particularly the cost of financing the project. The implication is that landholders in South Africa will respond unenthusiastically to the notion of making land available for the poor and, if they are required to set such land aside, it might well be that previously profitable developments become unprofitable. A lower rate of land development would result in a higher cost of urban land, so the policy could well backfire.

LAND BANKING

'Land banking usually refers either to advance acquisition of sites for government use or to large-scale public ownership of undeveloped land for future urban use' (Shoup 1983: 146). The main reason for the first form of land banking is to prevent the premature commitment of too large a proportion of the land for private use – land for schools, clinics and parks is obviously much cheaper if it is obtained in advance of urban development. In the case of hospitals and universities, which are much more space extensive, the acquisition of land prior to urban development may, in fact, be a precondition to the very establishment of the institution. The second form of land banking may have two purposes: to enable the government to recoup the costs of installing services (in effect, a form of betterment tax); and to obtain land for urban growth, most often to o'btain land at its agricultural value in order to facilitate public sector housing efforts.

There is considerable enthusiasm for land banking. Where housing is concerned, the view of the United Nations (1983: 35) has been that the 'fundamental need is for Governments to obtain a sufficient degree of command of the land-supply market to ensure a continuous throughput of well-located land accessible to the poor'. Land banking, where the government acts 'as an intermediary in the land-delivery process', is considered to be a prime example of how governments can intervene in the market. With the advantages of land banking being so obvious, it is worth looking in some detail at why it is not more widespread.

In the first case, seemingly a common, but mistaken, image of how a LandCorp should operate is provided by the Venter Commission. It favoured the New Town experience of England and France and also the Mitchells Plain development outside Cape Town, which consists of a large dormitory suburb for about 280,000 people, about 26 km outside the city. Mitchells Plain is a good example of deconcentrated urbanization, with a large, racially zoned township being created at a great distance from employment and social and commercial opportunities. In other words, the Commission's image of how best to solve the housing crisis was to undertake large-scale developments outside the city, on green field sites. In addition to its racial character, this vision is extremely problematical because increasing urban densities and infill are not even considered. There is the presumption that urban expansion onto green field land is the manner in which cities should grow. The presumption derives from the supposed efficiencies of large-scale developments undertaken by a number of specially created development corporations. As a matter of fact, Europe's New Town experience is misinterpreted. They represented the creation of dormitory towns whose expense and failure to develop their own economic base have largely led to their abandonment.

The Venter Commission model exposes a fundamental dilemma concerning the characteristics of the land that is acquired. Land is cheaper the further out it is and, given financial constraints, it is to such land that the institution may be drawn. But distant, cheaper land is unlikely to serve the needs of the poor, especially with respect to convenient access to employment; and if the land is especially far out, in anticipation of future need, it is difficult to judge which distant locations are optimal – it may be that urban development does not extend in the direction of public land holdings. Thus the contrary pressure is to buy land on the immediate urban fringe, for here one can immediately address known problems and be relatively confident that the land acquired will be reasonably well located. The dilemma is compounded by the fact that well-located land will already reflect price increases and may be too expensive for public housing purposes. Over time, of course, the public agency should have acquired a portfolio of land holdings, with the close-by land having been acquired earlier when it was in agricultural use, but it is very unlikely that the agency, at the outset, would be able immediately to pinpoint and acquire such land.

The cost of the land is critical for any public agency undertaking land banking. This is because:

● the longer the agency holds onto the land, the more expensive it becomes; that is, while the price originally paid for the land may seem small, in reality the same money, wisely invested, might be worth more than the land would cost were it to be acquired at a later date;

● bureaucratic delays in preparing a site plan, undertaking the necessary investments in infrastructure and allocating the land can worsen the shortage of land;

● while the land bank may ostensibly be established with a view to addressing the needs of the poor, the land may not be allocated fairly; and

● ironically, 'Inasmuch as large-scale acquisition of public land increases demand and restricts supply, land prices tend to be forced up, hurting the urban poor in particular' (United Nations 1983: 42).

While it may seem that this is a no-win situation, there are various means of dealing with cost. For instance, it was reported in 1983 that the Burmese authorities were empowered to purchase land at prices prevailing in 1948 (ibid: 141)! Swedish authorities pay compensation on prices prevailing ten years earlier and French authorities use the level prevailing one year previously. Legislation of this sort, of course, is a reflection of the relative balance of power between landowners and other interest groups.

There are still a number of other problems with land banking. First, it seems that 'there are two essential factors for a successful programme of land banking: close connection with the planning and land-allocating authorities and coordination between the various agencies involved in development plans' (ibid: 42). It would be incorrect to assume that the fact of the South African government's becoming democratic would compensate for an administration seldom distinguished by its competence. Secondly, there is considerable need for an honest appreciation of deserving locations and secrecy during the process of acquisition. Failing this, the land of influential persons might have to be bought at exaggerated prices and/or they will be informed so that they can earlier acquire the land in order to be able to sell it advantageously to the government. Many inferences in the South African press lead one to suspect that practices of this sort have been present for some time. The outcome is an income transfer to the wealthy. The same maldistribution is achieved in a different way in Delhi where, through acquiring rural land cheaply and allocating it largely to more wealthy urban groups, the Delhi Development Authority effectively promotes a redistribution of wealth from poor farmer to better-off urbanite.

Thirdly, the land banking institution develops interest of its own. The Delhi Development Authority, surely an example of how not to engage in land banking, again provides the example. It acquired land for public use and to act as a land bank, its function being to service, develop and allocate that land. Funds were to be raised by selling leases, with a view to enabling it to acquire and develop additional land. The underlying principle was that the Authority would acquire cheap, agricultural land on the urban periphery and be in a position to prevent speculation on land prices. It was anticipated that this would allow increases in value to accrue to the Authority, that it would enable the city to direct future development through its control of fringe land and that it would ensure equitable outcomes. The original intention, not realized, was that serviced plots were to be allocated in a ratio of 50:30:20 to low-, middle- and high-income groups. (The actual distribution of low-, middle- and high-income groups was 76:21:3.) The Authority appears to embody the motives underlying this book but, over time, its main motive became one of increasing its own revenue. It exploited its monopolistic control of the land market by speculating in land and, through staged releases of land, actually drove up the price of land. Once the Authority had acquired a portfolio of distant and proximate land, it was in its interests that the latter land have a high price, thereby enabling it to acquire more capital and so to obtain more land. The lesson should not be forgotten that a monopolistic public sector authority is able to speculate in a manner that exceeds the capacity of individual private landowners.

Finally, it is a worldwide phenomenon that public land has been the primary target of squatter invasions, and public authorities may be unwilling or unable to regain control of their land. Where the government fails to recapture some or all of the costs of acquiring the land, then the ability to sustain a programme of land banking is obviously threatened. Furthermore, unplanned squatter settlement on public land will considerably increase the cost of any subsequent attempts to service the settlement.

It is somewhat disappointing, therefore, to refer to the recommendations of a recent United Nations panel of experts (Rodwin and Sanyal 1987: 21).

> A number of difficulties . . . make our group reluctant to recommend this policy except when the circumstances are clearly exceptional. To be successful, land banking requires administrative capabilities which many municipal governments in the [Third World] do not have. . . Land purchases also presuppose ample funds at low interest rates. Just the opposite is now the case. There are also the uncertainties of advance planning, land management, and price setting, not to mention the increased risk of squatter invasion which can be resisted only at some political cost.

CONCLUSION

Neither land readjustment nor land banking should be recommended for the PWV. Land readjustment appears to be superfluous in that its value has largely been found in circumstances where there are many small landowners, which is not the situation on the fringe of the urban centres comprising the PWV. Moreover, the purpose of land readjustment – extending the supply of land and paying for the installation of services – can be realized by other means. The policy recommendations that follow point to another manner of achieving the same goal.

The success of land banking is too dependent on a competent and honest administration to take seriously. Rodwin and Sanyal (1987) specified that the various government agencies undertaking development plans, town planning and land allocation had to be closely coordinated. At the same time, the land banking functions had to be undertaken secretly; otherwise there was tremendous potential for corruption. It is difficult to imagine that these conditions obtain now or that they will be present in the future. In addition, owing to current housing shortages and the 'migration backlog' resulting from the apartheid legislation, the problem for land and housing is more one of fighting fires now than of using considerable resources to accumulate land for the future.

The discussion of land banking presented a picture of high intentions followed by a dismal record. Nevertheless, there are many reasons for the various urban areas comprising the PWV having, or sharing, a LandCorp. The obvious one is that a LandCorp could be a potent agent in increasing the supply of serviced land. Another is that the number of landowners on the urban fringe is rather limited. A few farmers, agricultural smallholders and some mining companies constitute a rather limited number of actors. They are in a position to coordinate the withholding of land and to speculate, which would lead to high prices and a disjointed urban growth as suburban development leapfrogged recalcitrant landowners. A LandCorp with forceful powers of 'eminent domain' could act to ensure a preferable urban structure. This means that it would be involved not solely with the needs of the poor. It could represent the public in the market for land and acquire speculative land holdings or holdings that, as a result of being withheld from the market, were leading to an efficient urban structure. A LandCorp would also serve to back up the threat of acquisition when vacant land taxes were introduced and landowners were asked to put a tax value to their land in a higher-value use.

These arguments suggest that, if it is possible to constitute a LandCorp in such a manner that it avoids many of the problems described in respect of the Delhi Development Authority, then a LandCorp might be worth trying. The LandCorp I propose would not be self-financing – it should have no interest in holding onto land (land banking) or in speculating in the land market. It would acquire land for the purpose of specific development projects and, because of the debt incurred when acquiring the land, would turn the land over as rapidly as possible. The LandCorp would obtain the necessary capital from commercial banks or a central development bank, and the income arising from projects would accrue to the local authority via the LandCorp. Local authorities would have responsibility for servicing the debts of the LandCorp and therefore would have a veto over the projects the LandCorp undertakes. In addition, the incomes of LandCorp personnel would be paid by the local authority. The LandCorp would, in effect, operate as the development arm of the city government and be subject to the direction of that government. This institutional arrangement is important if the LandCorp is not to become an institution concerned primarily with its own material gain and if it is to be subject to direction by democratic institutions. Under these circumstances, if local politics is democratic, there is every possibility that the LandCorp would retain a focus on the needs of the poor.

The beginnings of a land policy

Much of this chapter reads like a chronicle of good ideas that, in practice, have had unsatisfactory results. It is with some difficulty, therefore, that I have attempted to put together a set of recommendations that might constitute a land policy for the poor. It helps to make sense of the recommendations if they start at the national scale and then proceed to policy at the local level.

In 1980, 2.5 per cent of South Africa's GDP arose from the residential sector. Ordinarily the share of housing in GDP varies between extremes of 1 and 7 per cent, with the higher measure being characteristic of middle-income countries with growing economies (Renaud 1987). South Africa's measure is therefore typical of much poorer countries. What would the consequences be for housing if South Africa adopted South Korea's target of 6 per cent of the GDP coming from the housing sector?

In 1980, the public and private sector together built 54,557 dwelling units for all races in South Africa (De Vos 1987). These were generally built to a much higher standard than the minimal formal unit of R20,000. Nevertheless, in a straight comparison, if 2.5 per cent of the GDP produced 54,557 units, 6 per cent might have produced 130,936 units. If one takes the high standard of construction into account, it also seems reasonable to suggest that, if the same resources were applied to the production of cheaper units, the output could increase to about 200,000 units. This is still far short of the South Korean goal of producing 5 million units between 1981 and 1991, but is nevertheless very encouraging – it far exceeds the supposed 1986 private sector maximum of 70,000 units. It might be objected that a target of 200,000 low-priced units per annum is therefore overly ambitious, but the point here is not to quibble with the figures; there is no pretence of precision. My intention is to obtain an order of magnitude of what might be possible with both rapid economic growth and favourable housing policy. If we compare the share of investment in residential buildings in South Africa, which, as a percentage of gross domestic fixed investment, declined from 17.0 per cent to 10.77 per cent between 1960 and 1980, but then increased to 15.1 per cent in 1988, then a norm for less developed and middle-income countries of 15–30 per cent (Renaud 1987) suggests that it is not unreasonable to expect major increases in output in the residential sector.

The poor showing of the private sector does not point to malfeasance on its part – it is driven by the need to make profits and a greater contribution by it to the housing market will only follow on fewer regulatory obstacles, an increase in the relative rate of return obtainable in the market and/or government subsidies. Prior to the Black Community Development Amendment Act of 1986, the private sector was discouraged from serving the black market. Now, owing to the 1986 Act, which encourages private enterprise in the black land and housing markets, the situation has reversed markedly – in 1984/5 the private sector built 1,503 units for blacks, in 1987/8 the figure was about 45,000 units.

Evidently the government can both hinder and help private sector housing production. For this reason it is worth remembering the Venter Commission's suggestions regarding 'Speeding up of administrative and other processes involved in township establishment . . . [and other] Methods of curbing costs . . , eg. [reducing] development standards and making [State-owned] land available'. These are all a *sine qua non* for an effective housing policy. For example, when a developer buys land and then has to haggle with the bureaucracy for a couple of years over township plans and standards, there are tremendous costs involved in financing the initial investment. An expedited approval process will reduce costs. So, too, will lesser standards and the diminished use of professionals. Reduced costs will enable the private sector to build cheaper houses. The private sector is still some way off from serving the R20,000 a unit market – the higher-income black market, which accumulated prior to the removal of influx control and the Black Community Development Amendment Act, has still to become saturated – but, once it is, competition will force construction companies to serve the larger but less profitable (per dwelling unit) lower-income market.

Should subsidies be added to the legislative reversal, will these help the lower-income market to obtain housing? It is difficult to believe that the private sector will ever reach the really poor. One cannot specify from De Vos' (1986) work on black income distribution where the really poor cut-off line is to be found. The World Bank has stated that it has never been able to reach the bottom 20 per cent of a Third World city's income earners with a housing programme (IBRD 1975). One might argue that, since South Africa is a middle-income country, the figure here may not be so depressing, but a recent World Bank study reports that in Tunisia, a middle-income country, fully 30 per cent of the urban poor could not even afford access to illegal shanties (cited by Renaud 1987). At any rate, a cut-off line between the poor and the really poor seems wholly arbitrary. Suffice it to say that there is a substantial black population who it is realistic to expect cannot obtain access to a house or a serviced site. It is impossible to escape the conclusion that a significant section of the population cannot rely on the private sector for housing and will, forever, be dependent on public subsidies (Boleat 1987).

But, having concluded that subsidies are necessary, it is still an open question whether they will ever be sufficient or effective. Socialist and Third World countries have typically devoted relatively few resources to housing. 'Housing is invariably low on a list of government priorities, and is usually treated as a welfare issue' (Moser and Peake 1987: 195). As long as there is a dispute among development economists about the contribution of housing to the economy, and as long as governments have higher priorities to do with the expansion of the economy and of employment levels, housing will receive little help. Housing is something governments choose to prioritize once income and employment levels are more satisfactory.

Another aspect to the subsidy question is that, public pronouncements aside, governments seldom try to deliver housing subsidies to the poor. The primary beneficiaries of 'subsidies through housing programs are found in the second tier of the housing market where the salaried, private, or civil service middle-class households are concentrated, (Renaud 1987: 189). While these are not the people who most need subsidies, they are often groups who are well organized in order to obtain public support. The reverse is true for the poor.

Conversely, the problem with actually subsidizing the poor is that the subsidy provided may not reflect their priorities. Surveys reveal that, as the household income declines, so that relative focus on, for example, cheap food, free schooling and health services increases. If the government did supply shelter at highly subsidized prices, or for free, and even if the resulting allocation actually was to the really poor, the dwelling units would rapidly be reallocated by downward raiding. The only way that the aid that the government offers the really poor will stay in their hands, is to supply them with conditions that will not prompt downward raiding.

It follows that probably the best way of reaching the poor is to deliver facilities that, on the whole, are unattractive to higher-income groups and will not prompt downward raiding. This may seem like a cynical justification of doing very little for the poor but, in fact, the reverse is true. While one provides minimal help per household, this costs less and the government is able to reach more households. What it leads to is an argument for a limited subsidy for the supply of land, but with a low level of services.

The problem with this conclusion is why the government should be concerned to address the housing needs of the poor. It was argued earlier that neither liberal nor Marxist parties have any intrinsic motivation to serve people in the low-paid formal sector, the unemployed and the informal sector. However, the situation may be different at the local level. On the one hand, local governments have a natural predisposition against accommodating the poor – they do not offer much potential for rates and taxes and, instead, they appear to represent a drain on resources. On the other hand, it is a common thesis, one borne out by progressive practice elsewhere in the world, that the poor are more likely to gain influence in local than in regional or national government (McCarthy and Smit 1984). This may occur through urban social movements, or through a democratic struggle where the votes of the poor are significant to the outcome. Thus, if there is to be public intervention in the supply of housing or serviced sites, it is best located at this level of government.

The consequence for land policy is to suggest that responsibility be left in the hands of the cities comprising the PWV. What should the local government do in order to enhance the access of the poor to land? A number of measures, both direct and indirect, appear worth investigating.

It is especially important here that one remembers that the issue is not simply one of increasing the supply of serviced land. The goals are to promote urban density, to ensure a more convenient relationship between residence and place of work and other facilities, and to restrain price increases. There are also a number of trade-offs. For example, instead of servicing expensive in-town land, more distant land coupled with a cheap transport system might serve the purpose. Similarly, the zoning of industrial land and the siting of residential schemes should occur in such a manner that it relieves the commuting burden for labour. These are fairly obvious points and they do not need much comment but they do contain a warning that one should not think of the supply of land as an isolated issue.

Two indirect measures have already been specified – a progressive site value tax system and vacant land taxes. Both should be directed at influencing development to occur in a more dense and spatially concentrated fashion. The additional revenue generated by the taxes should be used for providing bulk infrastructure. This is because the primary reason for high land prices is not the physical lack of land as such, but rather the slow increase in the supply of serviced land. If the residents of low-income developments have to pay for the services, the rate of addition to the supply of serviced land will be very slow, with the result that land prices will rise still further. The two land taxes will help to spread the burden of financing bulk infrastructure and will promote higher densities.

Local governments should also be vested with authority (without requiring approval by provincial government as is presently the case) for specifying building standards and for passing township development plans. Decentralization of this responsibility should expedite the planning procedure and will ensure that the standards are closer to the aspirations and means of the people.

This begins to look as if I am putting all my eggs in one basket. What if local governments remain unresponsive to the poor? One is reminded of the position of blacks in Alabama during the period when Kennedy initiated many human rights reforms. Had it not been for deliberate federal intrusion into Alabama's affairs, many reforms would have been lost. The lesson is that, despite hopes that they would represent the poor among their constituency, local governments should not be left unattended.

Probably the best means of influencing the form of development undertaken by local government is found in control of the finances. For example, in the face of ruinous commitments by homeland governments, the Development Bank of Southern Africa was partly established in order to review capital commitments from the homelands (and now increasingly from local governments) as if they were development projects. This means that the projects have to pass certain viability tests, including that of public participation. The World Bank has found that the participation of project recipients in decision-making is especially important in housing projects; otherwise cost recovery is seriously impaired (Baum and Tolbert 1985).

A development bank would also help the government to avert continuous calls for assistance from local authorities. The government could provide cheap capital to the bank, for the purpose of providing serviced land, and it would thereby have a constant measure of, and control over, the assistance it was providing. This is an important consideration, since information regarding the extent and form of subsidies is frequently buried in the system and protected by vested interests, with the result that the government can ascertain neither the cost nor the effectiveness of the subsidy.

What this means for local governments is that, in the hope of democratic outcomes, they would be left to set standards and spend the money they raise as was politically expedient; but the additional capital they needed in order to undertake large housing and infrastructure projects, should be obtained through an institution such as the development bank. The bank would be in a position to require a targeting element in favour of the poor and, also, participation by them in the design of the project.

The local authority vehicle that could be used to enhance, not monopolize, the supply of land, would be a LandCorp, which would be set up with a view to attempting an adequate and timely supply of reasonably priced land. It would have the right to acquire, develop and sell land, all the while attempting to earn a profit on land sold to the rich so as to be able to provide a cross-subsidy for land occupied by the poor. However, below-market prices always result in excess demand and create a rationing problem, which itself works against 'more marginal' households that lack political connections. In this respect there seems no fairer means of allocating subsidized, serviced land than a means test, followed by allocation based on a first-come, first-served basis.

No legislation, however, has any chance of success if its bene-ficiaries are solely the poor – a wider constituency has to be sought if the LandCorp is to get off the ground. In this regard, the majority of the black population face financial problems and the LandCorp should be presented as a state-directed attempt to increase the supply of land and to restrain prices. This is exactly what it should do. The LandCorp might also be articulated as a government attempt to liaise with the private sector, and coordinated development projects may well be worth considering. An obvious example of this would occur when the LandCorp provides the land and the private sector develops part of that land for housing.

At no stage, however, should the LandCorp attempt to monopo-lize the land market. Instead, the private sector, including coopera-tive groups and employer concerns, should be urged to participate in the housing market. The greater the number of actors and the more diverse their goals, the more likely it is that people's needs will be served. Any attempt to deny profits to the private sector will ensure that the rate of housing construction is considerably diminished. Self-initiated activity in the middle- and upper-income housing market will serve a number of needs – the supply of hous-ing, employment creation, a market for backward-linked industries – and set in place a filtering process where the existing housing stock is made available to the less wealthy.

In conclusion, measures that are intended to promote increased density should be left in the hands of the local authorities. These include vacant land taxes, a progressive site value tax system and whatever associated changes in zoning and other policies are appropriate. The local authority should also be encouraged to become a more forceful actor in the supply of land by way of a LandCorp, which can be used to counter the small number of land-owners, and collusion in the release of land and land speculation, as well as being employed as a means of undertaking numerous extensions to the supply of serviced land. Distributional discipline for local authorities will be implemented via a central development bank, which will provide the LandCorps with access to cheap capital, provided that their undertakings meet criteria that reflect the interests of the poor.

Notes

1   This was the figure suggested by Jill Strellitz of the Urban Foundation during her address to the Association of Building Societies, 10 May 1988.

2   This estimate was provided by Roger Boden of the Department of Town and Regional Planning, University of the Witwatersrand.

3   The Household Subsistence Level is an estimate of the minimum income needed by a household to survive over the short term. It includes food, fuel, clothing, lighting, washing and cleaning materials, rent and worker's transport. It excludes health care, recreation, other transport, the purchase of household equipment and other essentials. The rent included above is set at a minimal level, considerably below the cost of entering the private sector market for land and housing. The Household Subsistence Level is usually contrasted with the Household Effective Level, which is one and a half times the Household Subsistence Level. The Household Effective Level reflects the fact that most households will devote a third of their cash income to items other than those specified on the Subsistence Level.

4   Downward raiding refers to a process where middle-income families, themselves poorly housed, acquire the housing intended for low-income groups.

5   Mentioned by Professor W. Doebele during his course on the 'Implementation of Metropolitan Planning in Developing Countries' at the Graduate School of Design, Harvard University.

6   At a luncheon seminar of the SPURS group, Department of Urban Studies and Planning, Massachusetts Institute of Technology, December 1986.

7   The mining land on the map depicts a combination of actual mining works and mine-owned land. The problem with the mining land is the lack of information. Some mining houses are reluctant to make information available regarding where their activities are, where they have land, and where they have leased land. Thus, when one combines a dated map, other specific sources and aerial photographs that were flown in 1987, one gets different results. The map should therefore be viewed as hypothetical. In respect of density, all built-up areas were marked using aerial photographs, and the densities were calculated using a 1:50,000 topographic map and a 1:10,000 street map, which showed individual plots. A problem that arose was that some of the built-up areas were not located on the street map, and in these cases my assistant relied on the 1:50,000 map. However, certain recently developed areas were not indicated on either map. In this case, if the area represented an extension of a low-density area, it was assumed to have the same characteristics, and so on. In general, though, it should be understood that Figure

4.1 is not a map but a relatively precise sketch of land uses and densities.

8   Mr M. C. O'Dowd first brought this point to my attention.

9   The section on taxation makes extensive use of an article by Smith (1979). Rather than repeatedly reference him, a concern for style suggests simply that I acknowledge my extensive debt to him.

10 These advantages become obvious when one undertakes a comparison with the other two alternative tax systems. In the case of the annual value or rental system, the rate is assessed against the annual rental value of the property (land and building). In the capital value system, the property tax rate is assessed against some proportion of the market value of the property. It is clear that both these systems have the effect of increasing the tax burden when developments or improvements occur. In the site value system, the rate is determined by the location of the land, its access to infrastructure, and intrinsic qualities such as the land's slope. Development on individual plots or tracts of land does not have the effect of increasing taxes. It is urban growth, public investment in infrastructure and the development of the area that increase the value of land in the area.

11 Tax capitalization may be explained as follows. The price of land, P_1, equals the present discounted value of the future yearly stream of earnings arising from the land.

$$P_1 = R_i/(1+r)^i$$

R_i = the value of gross earnings arising from the land each year. Because land does not depreciate, the revenue arising from land is assumed to occur for an infinite number of years.
r = the appropriate discount rate
i = $1 \ldots n$, number of years

Both the revenue arising from the land and the discount rate will change over time, and P_1 reflects landowner and buyer expectations regarding these changes. If one assumes, for illustrative purposes, that R_i and r are constant, then

$$P_1 = R/r.$$

The effect of tax capitalization is now easily demonstrated. If a tax, T, is imposed on earnings arising from the land, and the tax is expected to continue indefinitely, the present discounted value of the tax liability is T/r. 'Since a prospective buyer of land will visualise the stream of benefits from the land on a net-of-tax basis, he will internalise, or capitalise, the tax into the price he is willing to bid for the land' (Grimes 1977, 363). Thus:

$$P_1 = R/r - T/r \text{ or } = (R-T)/r.$$

5 Regional economic policy

By now, it is clear that policies intended to divert growth from the PWV are neither desirable nor workable. Nevertheless, it is probable that regional policy will be employed in a democratic South Africa in order both to restrain the growth of the PWV and to reduce spatial differences in income and employment levels. Because of this, I have devoted the first part of this chapter to a discussion of the likely political origins of regional policy.

Since the position advanced below is that economic activity and migration largely cannot, and also should not, be directed to other centres, much of this chapter urges against ambitious regional policy. However, my view is that regional and ethnic political competition will mean that regional policy is inevitable in a democratic South Africa; thus the chapter continues with an examination of the theory and practice of the most commonly employed regional policies. These policies involve: growth centres and industrial decentralization; industrial deconcentration; and, most recently, a secondary city strategy. Experience reveals that the decentralization policy has seldom been successful, but that industrial deconcentration is less flawed, having been effective in a number of instances. Secondary city strategies are still relatively new; they represent a theoretical advance on growth centre policy, but are still unproven. These policies, as well as the socialist experience, are described in the second part of the chapter.

South Africa's regional policy has been particularly inept and insupportably expensive. Aside from its apartheid context, however, harsh criticism of the effectiveness of the South African policy is probably unjustified; since decentralization policy has achieved so little success elsewhere, it might well be that one is criticizing a policy for being ineptly implemented when, in fact, it never had any chance of succeeding anyway.

It is this poor record of regional policy that causes me to question whether it is feasible to stimulate investment in lagging regions and to direct growth away from the PWV. The two policies have historically been closely linked since the means of promoting growth in the periphery was usually taken to require that investment be diverted from the core. This position, however, is less prominent nowadays and secondary city strategies represent an attempt to separate out the fortunes of core and periphery. I do not believe that the goal of slowing the growth of the core is realizable – the growth of large cities has not been slowed by regional policy (Rodwin and Sanyal 1987) – but this should not be taken to mean that the economies of lagging regions cannot benefit from public policy. I will present a sceptical evaluation of both the need for the government to redirect investment from the PWV, and its ability to do so; and I conclude the chapter with the recommendation of a secondary city strategy. The scenario shows how the government can enhance the development prospects of larger centres away from the nations core.

The political origins of regional policy

Throughout the world, within every constitutional and economic arrangement, governments attempt to influence the location of economic activity and population. While commonly cloaked in the faded rhetoric of cities being 'too large' or there being an 'optimal city size', the origins of regional policy are, invariably, political and, while there are differences between First and Third World countries, they are only differences of degree. In the First World, democratic processess cause regional disparities to become political issues. A government that fails to deliver employment and services to a poor region may expect its candidates to fare poorly there at the next election. In contrast, a Third World government, in addition to dealing with regional inequality, has to face elite alarm at the explosive growth of the urban poor. This has led to policies intended to slow urban growth and to restrain rural–urban migration.

Similar motivations appear likely in a democratic South Africa, and it seems that one should expect policy directed at the growth of employment in certain urban centres and regions rather than others. A reason for this is that spatial inequality is synonymous with differences in the economic and social opportunities available to competing ethnic groups. 'Tribalism' is a dirty word and has long been despised in South Africa because it has been employed both to divide the black population and to unite the Afrikaners. But, with the advent of democracy, should one expect that South Africans will differentiate between class-based and ethnically divisive appeals in the struggle for resources and that they will resist the latter? For example, should we expect that the Zulus in Natal, where the per capita gross geographic product is only 56 per cent of the national average (Smith and Coetzee 1987: Table 2.1), will accept their economically deprived situation relative to the rest of the country? Since the Zulus are a majority ethnic group, many will see considerable advantage in playing the divisive ethnic game in the pursuit of an improved allocation of resources.

None the less, progressive South Africans, no doubt in response to the National Party's continued 'strong emphasis on group rights',[1] are probably too quick to deride discussions surrounding ethnicity. Ronen (1986: 6, 7) illustrates the complexity of the issue:

> Numerous studies of European ethnicity have shown that it is not poverty but prospects for advancement that enhance the utilization of ethnic identity. The case in the 'developing world' has been similar.

> Long-existing ethnic identities could be put on the back burner for an indefinite period; but when the political struggles started, the ethnic identity could be used for political purposes.

The arrival of majority rule will bring with it just such a prospect for advancement, and ethnic identities, disparaged during the liberation struggle, are likely to assert themselves ever more strongly and to give rise to competition among regions.
But is this 'bad'?

> [The] reemergence and the reassertion of ethnicity in the mid-1960s may also be seen as a reaffirmation of a long-existing ethnic identity *in* the process of positive development – as an *integral part* of development where the state (or at least certain aspects of it), not ethnicity, is an obstacle to development. . . . the always available ethnic identity presented itself as a convenient rallying point to be utilized as a political instrument for development gains. . . . [Ethnicity] is no more disintegrative than political . . . party formation.   (Ronen 1986: 6)

Ronen's view that it might well be a positive move to encourage ethnic identity – since such identities provide, 'at least temporarily, the societal needs that eventually the state or nation is to provide' (p. 7) – has little currency in South Africa. None the less, Buthelezi's Inkatha and its like aside, it does seem that there should not be a prior presumption that ethnicity is a negative force. If ethnicity is the vehicle that enables poverty-stricken regions to extract resources from the state, then ethnicity should not be scorned.

A further reason for anticipating regional policy is that rapid urban growth is likely to elicit the typical policies intended to relieve growth pressures on the PWV. International comparisons of the behaviour of recent migrants reveal that a legitimate government would have no need to feel threatened by a rapid migration of poor into the PWV (Bienen 1984). However, it is instructive to consider that Zimbabwe, under Mugabe, has adopted an authoritarian attitude to squatter camps. The analagous situation in South Africa – where the ANC's primary constituency, the petit bourgeoisie and organized labour, will be in competition with the urban poor for housing, land and services and for jobs at the lower end of the scale – is that many would be expected to offer support for a policy directed at diverting future growth from the country's larger cities.

From growth centres to secondary cities

As I have mentioned, worldwide there are three especially common types of decentralization policy: industrial decentralization per se, industrial deconcentration, and a secondary city strategy. Policy appears to be moving in emphasis from decentralization and deconcentration to secondary cities.

Growth centres and industrial decentralization

South African policy up until now has predominantly consisted of industrial decentralization to distant locations. The critical difference between decentralization and deconcentration hinges on the distance of the site from a major urban area. Deconcentration points benefit from proximity to the city and access to large and differentiated markets, suppliers, services and labour. Decentralization points stand alone, and firms that are established in these areas find they must depend on the limited facilities available there, such as services, material inputs and skills. Alternatively they have to import them from elsewhere in the country and/or from overseas; create their own; or, through increasing the level of demand for those services or inputs, engender local development of new enterprises that service this demand. Although the last response is the theoretical optimum, the most common outcome of decentralization is the regional importation of goods and services.

Decentralization points have been rationalized through growth centre theory internationally and also, in South Africa, since the

1975 National Physical Development Plan. This theory holds that a growth centre is relatively large, rapidly growing and highly interlinked with the rest of the regional economy. The centre's growth is promoted through public investment in infrastructure, schools, health facilities and other social services as well as the decentralization of public offices; there are also a wide variety of incentives to the private sector to locate there. While the assumption has usually been that the investment to be targeted is that of large national and multinational concerns, this has changed, over time, to a relative focus on attempts to stimulate enterprise within the region.

As a result of the centre's growth, secondary effects are supposed to spread through the rest of the regional economy, thereby promoting regional development, but there are a number of implicit assumptions. One is that the centre selected has the potential to attract industry or services. Another is that the industry or services attracted will be large and highly interlinked, which is necessary if secondary effects are to occur. Lastly, it is important that the linkage effects occur locally. Failing this, the growth impacts of investment in a decentralization point follow the firm's pre-existing linkage network, usually to the major urban centres.

It was early on in the application of growth centre policy that planners realized the importance of focusing on centres with development potential. Development potential was usually reckoned in terms of an existing growth record, economic diversity, reason to anticipate future growth, an entrepreneurial base and a minimum, but unspecified, size for the centre. Without these conditions, there is unlikely to be much investment, and the same is true with respect to the formation of regional markets large enough for local enterprises to come into being. It was also realized that only a small proportion of a country's net capital formation in any given year involves locational decisions, and, of this, only a small amount is amenable to major geographic shifts. In America, for instance, only about 20 per cent of industrial relocations move more than 20 miles and this involves only about 0.5 per cent of the manufacturing employment in an area in any one year (Meyer, Schmenner and Meyer 1980). In other words, the investment available to decentralized locations is limited and, optimally, the decentralization that does occur should be directed to just a few sites.

The concern to limit the number of sites lies in the belief that growth centres have to achieve a 'critical mass' before growth within them can become 'self-sustaining'. If the investment is not focused by policy, it is held that the investment will be too dispersed for 'self-sustaining' growth centres to arise. It is now taken for granted that a country can sustain only a few centres, certainly fewer than six, without such points remaining eternally dependent on subsidies.

Internationally, the practice of growth centre policy has usually met with little success. One reason is that the centres were selected in terms of political criteria and it was difficult to resist the pressure to designate ever more sites. In addition, the investment attracted to the centres commonly consisted of branch plants of larger firms that had pre-existing input–output relationships. Their investment in a peripheral centre did not denote any willingness to utilize local suppliers, if, indeed, they even existed.

There is also the extremely important point that it is senseless to engage in an industrial location policy without prior regard for the spatial effects of the more general national industrial policy. For example, South Africa's import substitution policy leads to the concentration of industry where the market is, in the PWV; this outcome overrides official views on an optimum spatial distribution of industry.

Industrial deconcentration

Deconcentration refers to the location of production activities about an hour's drive away from the city centre. A recent advertisement for a deconcentration point north of Pretoria read, 'Ekandustria . . . is only 30 minutes from Pretoria and 45 minutes from Johannesburg by highway. (Which is so close you could say it's on the outskirts of town rather than in the outback of the country.)' It is for this reason that the process has been called the suburbanization of industry. If deconcentration points are well selected relative to labour supplies and transport infrastructure, they may prove attractive for industrialists who do not require frequent contact with suppliers and the market.

Deconcentration strategies, in one form or another, are found worldwide and are common when the concern is with metropolitan form, congestion, pollution and urban efficiency. The best-known response has been new towns on the metropolitan periphery, but these have proven very expensive and, as a policy tool, have been abandoned by middle- and lower-income countries. Even in the First World they never succeeded in becoming more than dormitory towns and, as it turns out, they did not displace the pressures on the city. London's attempt, for example, to create a green belt around the city between the old city and new development outside was overwhelmed by urban growth pressures.

Moreover, stringent restrictions on industrial expansion in cities may prove positively harmful. World Bank researchers note that '*Small industries come up in city centres as large ones move out. . .* small firms tend to locate in central areas [even though] land prices are high. They do not need much space. More important for them is access to markets, services and workers. This access is best in city centres, which act as incubators for small, new firms' (*Research Brief*, 1986, 7, 1; emphasis in original).

In retrospect, it seems obvious that, for economies that are slowly increasing the proportion of their labour force in the tertiary sector, an industrial (and associated residential) deconcentration policy can never seek to do more than redirect a small portion of the jobs created from more central locations. Also, in general, industrial deconcentration is a process that occurs naturally. For example, 'large establishments need a lot of land at a low price for plant expansion. It is more important for them to achieve economies of scale in production than to be near city centres. So, they tend to move out of central areas where land is scarce and expensive' (*Research Brief*, 1986, 7, 1). But this is a market-induced process, which occurs as cities and enterprises increase in size. This is typically held to occur as cities grow, as urban diseconomies become more consequential. Under such circumstances, urban designers may justifiably attempt to redirect future growth to alternative locations.

Will industrial deconcentration contribute to the efficient and equitable operation of the city and, in particular, will it facilitate the access of the poor to reasonably priced land and reduce their commuting burden? It is evident that the issue here is different from that of industrial decentralization. It is also clear that the issue is not one of national policy initiatives, but one of modest local attempts to intervene in market processes: deconcentration policy has less to do with diverting urbanization than with attempting to influence the form of metropolitan growth. In other words, deconcentration strategies are more the province of urban designers and local politics than of distant planners pursuing national urbanization objectives. These strategies, motivated by conceptions of the apartheid city, are prominent in South Africa but, since South Africa's cities are so small, there are still few grounds for treating deconcentration as a major issue.

Secondary cities

A secondary city is defined in terms of the national urban hierarchy and consists of the cities that fall in size immediately below a country's metropolitan areas. They are, as a result, larger in scale than the sites hitherto selected as growth centres. Secondary cities are usually thought of as having populations ranging from 100,000 to 2.5 million persons (Rondinelli 1983), but Richardson (1987) concedes that in Sub-Saharan Africa smaller centres of 20,000 persons may be all there is. The cities are dominated by commercial and service activities, not especially of a rural or agricultural nature, and they will also have a relatively high proportion of their labour force in the informal sector. In addition, compared with the metropolitan areas and the proportion of the latter's labour force in the manufacturing sector, they have a smaller share of manufacturing activities and employment. Due to their size, and the fact that they already contain a fairly diverse economic base, as well as the fact that only a few centres are selected, they are deemed to have the greatest potential for development.

In Richardson's (1987: 213) view, a secondary city strategy

stresses indigenous development, such as agro-processing, small scale industry, and the informal sector, rather than the attraction of large-scale industry from outside; it builds on potential interactions between the growth of the city and the agricultural character of its hinterland; and it gives relatively more attention to social infrastructure than to an almost exclusive reliance on economic and industrial infrastructure. This last point suggests that a shelter and services investment program will be a key component of a . . . strategy.

The emphasis he puts on shelter and social services is extremely interesting since it suggests a potential marriage between a secondary city strategy and an attempt to exploit the benefits gained from public expenditure on the provision of serviced sites for housing. The local, preferably small-scale, construction industry and, also, industries serving the market for household durables should see many growth opportunities. The government can, for example, choose to make a disproportionate allocation of expenditure on shelter to a few secondary cities.

More generally, Richardson shows the extent to which policy recommendations have moved away from growth centres and decentralization – the location subsidies of the past are done away with and investment in infrastructure is much more selective. Since a 'policy stressing the long-distance relocation of industries may be far less effective in promoting decentralized growth than one aimed at improving conditions for local economic development' (*Urban Edge*, 1986, 10, 7: 8), the focus is, instead, on enabling regional enterprises to help themselves by expanding their markets and developing local linkages. Growth in outlying centres originates not so much from industrial decentralization as 'from local firms that are just starting up or that are expanding their local operations' (*Urban Edge*, 1986, 10, 7: 8). The fact that regional enterprise is usually smaller also helps to explain the move away from decentralization incentives, since industrial location policies typically strive to attract investment from large corporations. Small firms are tied to markets, suppliers, services and the spatial constraints of management's knowledge, and, commonly, are not amenable to major locational shifts.

Rondinelli (1983) is unwilling to abandon decentralization; he adds the option of limited decentralization to the above policies, as well as some decentralization of public offices. In the former case, where regions have specific comparative advantages, the inducements might be tailored to specific industrial sectors, so that both geographical and sectoral concentrations of industry occur. The desirability of such concentrations can be in relation to relative labour surpluses, natural resources, or access to transportation infrastructure.

The decentralization of public offices is especially appropriate when the nation has a federal constitutional structure. In relation to a secondary city strategy, federalism serves two purposes: it promotes the decentralization of public offices and expenditure and it is associated with certain attitudes on the part of regional elites. Rondinelli (1983: 182–3) believes that secondary cities are more effective sites for decentralization when

> local elites . . . identify their own success and status with the economic growth and social progress of the city and its region . . . [when they] invest their resources in the growth and development of the city rather than investing surpluses generated from city activities in other places . . . [when they] are innovative and aggressive in introducing more effective methods and techniques of production to increase output and income within the local economy . . . [and when they] are aggressive, and successful, in bringing external resources into the city.

In effect, when secondary cities are the capitals of federal states, the probable success of a regional strategy is enhanced.

Secondary cities probably represent the state of the art on the international consulting circuit, but it is still appropriate that the secondary city strategy be recommended with considerable restraint. It is generally accepted that it has historically most often been the case that regional inequality was reduced by economic development (Williamson 1965; Hall 1987), and that the inefficiencies introduced by 'premature' or poorly considered regional policy may well retard that development. Economic growth and employment creation in a region reflect national and international trends more than they do forces within the region. For example, policies intended to promote investment in, or decentralization to, a region that is characterized by cyclical unemployment and under-utilized industrial capacity will do less to relieve unemployment than will short-term measures designed to increase the level of effective demand.

Moreover, it is clear from the way this discussion has proceeded that the secondary city strategy is largely a response to the earlier failure of growth centre policy. It still represents a spatial strategy taken out of context. Spatial strategies simply have to follow on prior clarity regarding national industrial policy, which industrial sectors are growing, and the locational preferences implicit therein. It is important to bear in mind that 'implicit spatial policies more often than not conflict with explicit spatial policies with the result that efforts to redistribute population and economic activity are consistently undermined by the primate city[2] and core region biases inherent in many sectoral policies – import substitution, subsidized urban services, internal terms of trade distortions, etc.' (Richardson 1987: 209).

Regional policy in socialist countries

Regional policy under socialism has been more successful than that in Western and Third World countries, but not markedly so. As promised in Chapter 1, my portrayal of that policy is undertaken in order to ascertain whether, through considering policy in First and Third World countries, I have not missed possible alternatives.

Urban experiences under socialism are highly varied (Murray and Szelenyi 1984), but the common sentiment is against the growth of the large cities. This view was derived from Marx and Lenin's perceptions of an antagonism between town and country, founded on the capitalist division of labour and the relatively privileged position of industrial labour as compared to agricultural labour. Thus, 'With the liquidation of the class structure, and with the rebuilding of society along communist lines, the existing differences between town and country will gradually be eroded' (Khodzhaev and Khorev 1978: 512). Agricultural labour was to be transformed, with the labourers living in towns and working under conditions modelled on industrial practices. The means to this end were taken to include policies directed at a 'greater equalization of the distribution of large scale industry and of the population over the country, . . . and overcoming the excessive concentration of population in the large cities' (idem.).

Practices have varied from the vicious extreme of Cambodia; to 'labor books and internal passports [which] were attempts at controlling the pace and direction of migration streams' in the USSR (Huzinec 1978: 141); to the less onerous example of Budapest where, until 1965, access to housing was denied to newcomers (defined as those who had been there for fewer than five years). The effect in Eastern Europe and the USSR was to keep 'a significant fraction of the industrial working class . . . out of the urban centres, and away from the subsidised better urban infrastructure' (Murray and Szelenyi 1984: 99). Moscow's and Budapest's policies are decidedly reminiscent of various stages of 'influx control' under apartheid South Africa.

It seems that, as a result of 'comprehensive wage and price systems' and 'comprehensive education, health services and social policies', planners were able to slow the process of regional differentiation associated with industrial development in the developing world (Musil and Rysavy 1983: 523); and that they were also able to prevent the extremes of city size realized in capitalist countries (Murray and Szelenyi 1984). Nevertheless, there was still perceived to be 'excessive' industrial and urban concentration and the associated shortages of services and housing. The cause of the industrial concentration was that 'efforts to exploit available investment outlays in the economically most effective ways [led to] the concentration of economic activities in large cities and agglomerations' (Musil and Rysavy 1984: 523). In the Soviet Union, for example, the concentration was a result of the relative predominance of industrial ministries over regional planning agencies. The former had calculated 'that the productivity of labor in cities with a population over 1,000,000 is 38 per cent higher than in cities with a population between 100,000 and 200,000, while the return on invested capital is twice as high' (Huzinec 1978: 143). If plan quotas were to be realized, there was obviously good reason for this to be achieved by 'expanding existing facilities or by developing new plants in areas already providing support services' (ibid.: 144).

Thus, while the results in socialist and non-socialist countries were different, they were not so marked that Western planners would not empathize with the uncertain response in Hungary to urban migration and housing shortages. Urban planners in Budapest thought that the best way to restrain the growth of the capital was through a ring of satellite cities based on the Moscow Master Plan and the Greater London Plan. These cities were supposed to 'absorb' migrants coming to the capital.

This idea was rejected by the regional planners because it would have concentrated urban investment around the capital and they proposed 'counterpole cities'. The regional policy was initiated in the same year that the Budapest Master Plan was accepted (Hegedus and Tosics 1983).

Undoubtedly, the relevance of the socialist material is limited. The success of the policy has been contingent on the ability to direct the location of economic activity and a preparedness forcibly to control the location of population. The former is incompatible with a mixed economy and the latter assumes totalitarian powers, which, one hopes, characterize South Africa's past, not its future.

South Africa's decentralization policy

Many of the weaknesses of South Africa's industrial decentralization policy derive from its goals – the weaknesses are, in a sense, 'built in'. To start with, there are too many centres. There are 49 industrial development points (IDPs). Many of these are double-IDPs, where both a white town and also the adjacent black township across the homeland border have IDP status. The Decentralization Board also has deconcentration points, other industrial development points, *ad hoc* sites, . . . and has been prepared to consider subsidies at over 200 sites (see Map 2.2). An official in the Transkei told me that the whole area is, in effect, an IDP because reasonable applications, regardless of their location, are given favourable consideration. If there is considerable political commitment to averting black urbanization, and if homeland development corporations compete for investment, then it is impossible to employ decentralization benefits solely for the purpose of attracting 'propulsive' industries to a few centres with development potential, as prescribed by growth centre theory. The result has been the creation of minor industrial enclaves whose longevity appears to be determined by the period over which the respective industries are eligible for concessions.

A further consequence of inadequate monitoring of the types of industries that receive subsidies was that in 1986/7, for example, only 6.76 per cent of the applications for concessions were rejected (Decentralization Board, *Annual Report 1986/7*).[3] It is in the nature of things that bureaucrats charged with administering the decentralization scheme will measure success in terms of the number of jobs created or relocated, with little regard for the cost involved; and it is therefore no surprise that the survival of up to 60 per cent of decentralized employment is contingent on subsidies. It is also not surprising that the Decentralization Board fails to report on the rate of enterprise failure, which has informally been described as 'spectacular' by an accountant closely involved with the scheme.

The goal for the decentralization programme was specified by the Tomlinson Commission (South Africa 1955) as the creation of 20,000 manufacturing jobs and a further 30,000 jobs as a result of multiplier effects. Thirty years later, South Africa has about 320,000 blacks entering the labour force each year, about half of them in the homelands. The limited success of the decentralization policy is shown in Table 5.1, where the number of jobs claimed by the Decentralization Board seldom amounted to a tenth of 320,000 per annum. (The jobs attributed to the decentralization programme since 1982 are less than those indicated in the table as the table reflects approved projects and usually fewer than half these projects are actually implemented.)

Decentralization has had only a marginal impact on unemployment in the homelands and, at the same time, it has more than proportionately exacerbated unemployment in the rest of South

Table 5.1   Projects established, capital invested and employment created under the auspices of the Decentralization Board, 1975–1987[1]

Year	Projects established or expanded	Private sector capital investment (Rm)	Employment
1975	59	68	5,730
1976	91	88	6,738
1977	136	174	12,469
1978	123	112	8,408
1979	175	132	11,178
1980	152	140	11,421
1981	214	145	11,792
1982/3	774	2,459.8	63,729
1983/4	1,190	1,201.6	69,914
1984/5	1,216	1,176.7	77,486
1985/6	1,243	1,299.8	87,635
1986/7	1,027	1,262.7	68,780

[1] The data exclude the independent homelands. Up to, and including 1981, the reporting year was the relevant calendar year. Thereafter it is 1 April to 31 March of the following year. The new decentralization concessions package began in April 1982.

Source: Decentralization Board, *Annual Reports*.

Africa. It is preposterous that, in 1969, the Minister of Labour could boast (inaccurately) that, as a result of controls on the employment of blacks and the zoning of industrial land, the creation of 220,000 urban jobs had been prevented (cited in Gottschalk 1977: 51). Nevertheless, there is reason to be taken aback when Wellings and Black (1986) estimate that the cost of creating decentralized jobs is about four times the cost of equivalent jobs in the city. Since decentralization generally represents a move to low-wage regions, not only is employment creation forgone, so, too, is the black share of the national income.

The costs of the policy, as well as its inability to prompt regional development, are exacerbated by the selection of sites that sometimes have no pre-existing industrial infrastructure. Industrialists, themselves, complain about the lack of physical infrastructure and social services, and about having to pay management and skilled labour higher wages in order to attract them to forbidding locations (Addleson *et al.* 1985). Growth cannot be stimulated in underdeveloped regional economies through the introduction of a growth centre and a few industrial investments. The centres become industrial enclaves with all the linkages leaving the region. This last point is well illustrated in the following comparison of Isithebe and Pietermaritzburg. Isithebe is the most successful IDP in Natal, but Table 5.2 shows that linkages within the region are minimal, representing only 3–4 per cent of a firm's market, supplies and manufacturers. In contrast, were the investment to have been located in Pietermaritzburg, a secondary city in Natal, the linkage loss would be much smaller. One can see from Table 5.3 that instead of a firm obtaining 3 per cent of its inputs locally, it now obtains 20–30 per cent. In addition, instead of selling 3 per cent of its products locally, a firm now sells between 13 per cent and 40 per cent locally. It should be obvious that local linkage networks are, in fact, only

Table 5.2 Isithebe: location of inputs and outputs

Source	Markets %	Materials/ supplies %	Plant/ manufactures %	Services %
Local	3.7	3.8	3.1	8.5
Durban	18.5	30.8	13.5	75.0
Rest of South Africa	60.8	42.6	39.0	16.5
Overseas	17.0	22.8	44.5	—

Source: Wellings and Black (1986: 21).

Table 5.3   Pietermaritzburg: source of firms' inputs and location of markets

Source	Inputs (%)			Markets (%)		
	Raw Mat.	Proc. Prod.	Manu. Prod.	Raw Mat.	Proc. Prod.	Manu. Prod.
P'burg	30	29	20	13	40	33
KwaZulu/ Natal	20	25	22	43	32	27
Rest of South Africa	35	37	44	30	25	37
International	17	9	13	14	3	3

Source: McCormick and Associates (1986: 26 and 29).

possible in relatively large centres. If small centres are selected for decentralization purposes, then, regardless of the source of the investment, external rather than local linkages will inevitably follow.

If one adopts the bureaucrats' stance and evaluates decentralization solely through reference to jobs created, the policy's success should probably be divided into two periods. Following the announcement of the improved decentralization concessions, effective from April 1982, there has been a significant increase in the amount of decentralized activity. For example, between April 1982 and December 1985, the figures for industrial employment in most of the homelands increased by 60–95 per cent. Table 5.1 reflects this increase. The number of jobs decentralized each year increased from 11,792 in 1981 to about 32,000 in 1982/3 and reached a high of 43,000 in 1985/6. As explained earlier, typically slightly less than half of the anticipated investment actually occurred. The table is misleading because it excludes the independent homelands. However, although data for the independent homelands are less accessible, a similar rate of increase is suggested for these areas. For example, in June 1982, the Ciskei had 33 industries and, by December 1985, about 100 industries had been established there (Tomlinson 1986). The figures suggest that, while decentralized investment may have taken place without recourse to the Decentralization Board's concessions prior to 1982, thereafter the concessions became so attractive that industrialists could not ignore them.

In fact, Bell (1983, 1984) provides important data on an industrial decentralization trend, apparently not recognized by the Decentralization Board itself, in the pre-1982 period. The trend

Bell identified is that between 1965/6 and 1979 the growth rate of manufacturing employment in the 'main industrial areas' was consistently below, and that in the 'rest of South Africa' consistently above, the national average. The main industrial areas are the PWV and the Durban, Cape Town and Port Elizabeth metropolitan areas. (Between 1979 and 1982 the PWV grew more rapidly owing to the expansion of heavy industry.) The result was a changing share of manufacturing employment, where the relative balance shifted from 81.9 per cent and 18.1 per cent in 1965/6 to 76.9 per cent and 23.1 per cent in 1979. The actual increase in employment between 1965/6 and 1976 in the 'rest of South Africa' was 124,862 jobs. Although this is not a large number it is considerably more than the amount of employment attributed to the decentralization programme.

Bell relates trends in the location of manufacturing employment to growth cycles in the South African economy, which, in turn, are tied to international economic conditions and to a restructuring of the 'international division of labour'. In particular,

the observed tendency towards industrial decentralisation has occurred largely spontaneously in response to the pressures of competition in world markets for manufactured goods. The geographical deconcentration of industry has largely been the result of a striving after lower labour costs through the substitution of Black labour obtainable at low wage rates in the less industrialised areas, for more expensive White, Coloured, Asian and, indeed, also Black labour available in the larger industrial centres. There had always been large inter-regional wage differentials . . . , but the immediate cause of the intensified effort to take advantage of these was the structural change in the world system of industrial production and trade which occurred in the sixties (Bell 1984: 9–10).

In a nutshell, 'the changing inter-regional division of labour may be seen as a microcosm of the new international economic order' (Bell 1983: 63).

There are a number of reasons why one should exercise caution in accepting Bell's data and explanation, as Tomlinson and Addleson (1986) have enumerated. However, a reading of international experience and the relocation of labour-intensive industry from high- to low-wage regions and countries prevents any easy dismissal of Bell's position. The move from an international economy dominated by America to a highly competitive international environment – from 'monopoly to global capitalism' (Trachte and Ross 1985) – is impelling firms around the world to relocate and restructure their production, and also many of their service activities. While increased competition between firms and the resulting heightened struggle between capital and labour are translated variously by firms in different social and economic contexts, the general outcome has been a 'system of production and trade which has become more and more global in extent' (Dicken 1986: 3). If the world has seen a historic 'major transformation in industrial geography' (Massey 1984: 1), one has to expect some corresponding adjustments in South Africa.

However, in contrast to the international shift to global capitalism, and in part as a result of sanctions, in South Africa there has been an embedding of monopoly capitalism, with the dominance of the country's largest corporations becoming ever more pervasive. As I noted in Chapter 3, in 1987 the Anglo American group increased its controlling interest in companies listed on the Stock Exchange to 60.1 per cent, Sanlam controlled another 10.7 per cent and SA Mutual had 8.0 per cent. In addition, the government controlled a quarter of the assets of the country's top companies through the country's parastatals. The competitive pressures implicit in global capitalism are less prevalent: protected South African industries are less driven to seek cost–cutting measures and are also not obliged to confront the unions in the manner that has recently occurred in Western democracies. Quite the contrary, South African capital, prominently Anglo American, seeks to establish cosy relationships with the unions. This is possible in the industrial sector because of an environment of stable market share and administered prices.

Yet, at the same time, this does not refute Bell. Not all of South African industry is so concentrated. The textile and cloth-ing industry, for example, has led the industrial decentralization process. These industries have been particularly susceptible to foreign, mostly Far Eastern, competition and a common response has been the location of production activities at dispersed sites, where lower labour costs enable the firms to become more competitive. Thus, when textile industrialists left Johannesburg for Cape Town or Durban, they were moving towards established cheaper labour supplies and a sound infrastructure base. When, after 1982, the data suggested a large-scale move to Isithebe, Dimbaza and other locations supported by the Decentralization Board, industrialists were moving towards labour that, in some locations, was more than free. For example, my work in the Ciskei in 1985 revealed that unskilled women were commonly paid R45 per month and that unskilled men were paid R60 per month (Tomlinson 1986). The tax-free, quarterly cash grant paid to industrialists for employing this labour was R110 per month.

There is one factor, though, that complicates this assessment: peripheral labour is generally less productive than labour based in the PWV. Although wages are lower and subsidized in peripheral areas,[4] 'labour is considered a major "problem" for bantustan industries . . . The scarcity (and hence high cost) of skilled labour, and the low productivity of unskilled labour are frequent complaints . . .'. a surprisingly high proportion of industrialists in our survey considered labour costs (including managerial/professional categories) in decentralised areas to be equal or even higher than metropolitan areas' (Wellings and Black 1986: 18). In other words, it may well be that, when one takes productivity into account, lower wages may not necessarily imply cheaper labour.

The real motivation for decentralization, at least since 1982, appears to have been not so much cheap labour as the monetary concessions. Smith and Coetzee (1987: 40) report of KwaZulu, where 65 per cent of the industries were established after the introduction of the 1982 concessions, that '69.81 percent of the respondents indicated that they would not have decentralised if concessions were not provided. . .'. The concessions enable the companies to gain a competitive advantage.[5]

The marked increase in decentralization since 1982 might suggest that, whereas previously it was a failure, it is now a success. However, this conclusion is belied, first, by the cost of the programme. In the light of the fact that the direct subsidy cost of the programme is in the order of R729 million,[6] how can a policy affecting the location of about 30,000 jobs a year be deemed effective? One also cannot avoid being impressed by the fact that the cost of the transportation concession exceeds that of the wage concession – it amounted to R216 million (excluding the independent homelands) in 1986/7 and, unlike the wage concession, it is not removed after seven years. There is every reason to believe that the effect of the programme is to cajole industrialists into investing in inefficient locations. Another illuminating measure follows on the Ciskeian experience. Ciskeian planners estimated that the decentralization concessions per worker per year, calculated over a 10-year period, excluding capital costs, were costing R3,000 per annum. This estimate goes a long way towards explaining the homeland's rejection of many of the concessions and the introduction, in 1984, of a tax-free option as an alternative way of attracting industry. The tax-free legislation is actually expected to improve the Ciskeian government's fiscal situation!

Many of the above criticisms are now accepted in South Africa. The South African government is itself undertaking a review of the decentralization programme. From a planner's point of view, the point that most raises one's ire is the extent to which political motivations have overridden sound policy. In any country there is a tension between the pursuit of equality and efficiency and, also, between both these factors and the regional allocation of funds in order to win political support. In other words, political intrusion into sound policy is part of the game. But the South African situation is distinguished by the extent to which the political game has dominated good economic sense.

Does South Africa need a regional economic policy?

Is regional policy feasible?

Even if one were to decide that a regional problem does exist, there is not necessarily much that one can do about it. As Richardson (1987: 208) argues,

> A critical problem is that population distribution outcomes are the result of three sets of forces, the individual impact of which is difficult, perhaps impossible, to unravel. These forces are market trends and the dynamics of the aggregate development process; the implicit spatial impacts of macroeconomic and sectoral economic (and social) policies; and explicit spatial policies. . . . it is widely believed that explicit spatial policies are the weakest of the three sets of forces.

Spatial outcomes primarily reflect broader economic and political forces (Gore 1984). Kenya provides a good example. Urban growth there, especially that of Nairobi, resulted from economic policies that turned the internal terms of trade in favour of industry and against agriculture. The outcome arose from an import substitution industrial strategy that, owing to taxes on imported goods, allowed the price of manufactured goods to rise above those prevailing internationally, and from price controls on agricultural goods that set prices at, or below, those of the world market. These policies, in turn, revealed a distinct balance of power within Kenya (Tomlinson 1982, 1983b).

If regional policy represents an attempt to swim against the current, what is its effect? The answer, it seems, is not much. 'The conventional spatial plan – which designates this urban area for containment and that one for rapid growth – assumes a level of control over this process that is virtually unattainable in any system where there is a substantial level of economic autonomy' (Hamer, cited in *Urban Edge*, 1986, 10, 7: 2). 'Relocation incentives don't have much impact . . . Location policies can reduce efficiency. That is, location policies can reduce the efficiency of using scarce resources and thus lead to large welfare losses' (*Research Brief*, 1986, 7, 1). Thus, in South Africa, since seemingly up to 60 per cent of the decentralized employment is dependent on subsidies for survival, the implications for the economy are serious. This is because if a business has to be subsidized in order to survive, the value of its inputs, taken out of the economy, is greater than that of whatever it puts back into the economy. In order to sustain this situation, resources (and incomes) have to be drawn from other areas and reallocated to decentralized activities, leading to a corresponding decline in activity and employment in those areas from which the resources were drawn.

The impression created is, I am sure, that I am against regional policy. In fact, a trade-off underlies these arguments. One side looks to future unemployment and poverty. From this point of view, the evidence, such as it is, makes the future growth of the PWV appear desirable. 'There is a fundamental conflict between high economic growth and decentralization of population. If a high rate of economic growth is to be achieved, further concentration of population into a few large metropolitan areas cannot be avoided' (Mera 1973: 271; cited in Gilbert 1982b: 176). Thus, given 'the long-established association between increasing primacy and faster GNP growth and the more recently discovered positive links between the growth of large metropolitan areas and social indicators (school enrolments, literacy, declining infant mortality, nutritional intake, life expectancy, falling birth and death rates)' (Richardson 1987: 211), one should hesitate before advocating controls on the growth of large cities.

In respect of the PWV, this growth-oriented view finds strong support from the fact that the South African space-economy is sectorally differentiated. The PWV is the site of capital- and skill-intensive industry and this industry is not 'footloose'. Labour-intensive industry has, to a large extent, already left the PWV and, given these circumstances, Bell (1987b) has been led to ask what it is that a decentralization policy is supposed to decentralize!

The alternative view looks to the concentration of problems related to unemployment, transportation, housing and services in the large cities; the possible development of severe pollution, congestion and communication diseconomies; problems with administering cities; and regional inequality and the divisive politics to which it can lead. Many might feel 'skeptical about the benefits of unfettered urbanization which has resulted in the demand – impossible to satisfy – for employment, housing, transportation,

and other services . . . [Many might also] be concerned about
the growing regional inequalities exacerbated by the movement
of labor and capital towards the large urban centers' (Rodwin and
Sanyal 1987: 4). This point of view would be bolstered by the
position that regional policy is not so much ineffective as poorly
timed. That is, 'Spatial policies are much more likely to be effective
at intermediate stages of development when regional markets begin
to cross economy thresholds, pecuniary diseconomies and congestion costs may be emerging in the primate city and core region,
and polarization forces show signs of weakening spontaneously'
(Richardson 1987: 209).

Regional policy and economic development

The government claims to be motivated by two regional problems:
regional differences in income and employment levels and the
supposed need to divert growth from the PWV. While, under
apartheid, such claims represented thinly disguised attempts to
prevent black urbanization, a post-apartheid government will
elicit greater credibility when voicing such concerns. However,
both issues need clarification.

In the first instance, while regional inequalities are unfortunate, it
is unclear that development policy should reflect spatial constructs.
As Wellings (1985: 423) puts it,

> those studies which have advanced the analysis of regional
> inequality beyond description towards considerations of plan-
> ning strategy have always assumed that an improvement in
> regional equity is a valid and desirable policy objective.
> As a result, it is generally recommended that the state
> should intervene to allocate development funds according
> to regional needs. In practice, then, this means that the
> success of these development planning strategies is judged
> with reference to some yardstick of 'development' which is
> indexed to spatial constructs rather than social groups. The
> objective is therefore to raise the general level of development
> in impoverished regions as if the impact of state policy would
> be identical for all individuals of social groups within those
> regions. This is of course nonsense, what is missing from
> the analysis is some consideration of the social and political
> structures which determine how development expenditures
> are distributed.

Development which, say, raises per capita incomes in a region will
not benefit all people and classes equally. Thus, while policies

directed at relieving differences between regions have sometimes proven successful, they have more often been found to exacerbate inequality within the target or poor region (Stohr and Todtling 1978). The issue is clearly more one of addressing the needs of the poor than of evening out interregional averages.

The second issue arises from the belief that cities can become too large, or too large relative to other urban centres. With regard to the PWV's size, Table 5.4 presents a recent projection of the anticipated growth of the four major metropolitan areas between 1980 and 2010. The PWV comprises a number of relatively small cities which do not compare with Calcutta, Mexico City or Sao Paulo. It has been held that the scale of the latter cities has rendered them ungovernable. The PWV has a population growth rate of about 4.3 per cent and by 2010 will have become relatively large – although, again, not nearly as large as the expected size of other metropolises of middle-income countries. If the Witwatersrand retains its share of approximately 70 per cent of the PWV's population, it will consist of a sprawling urban region containing some 11 million persons. There is now widespread acceptance that for an urban region of this size the issue is more one of attempting to manage rather than to divert growth (Rodwin and Sanyal 1987). This will especially be true if South Africa achieves a decent economic growth rate, for then it will have the resources to manage the growth effectively.

The alternative question is whether the PWV is too large in relative terms. The question arises from the view that one city can be so much larger than the other urban centres that investment and growth are invariably located within it. From this it follows that a dispersed hierarchy of urban centres, while not creating 'spread effects', at least facilitates economic development in the periphery as the economy undergoes a process of growth and structural change. The PWV's economic dominance is clear from the point that about 43 per cent of the country's employment in the manufacturing sector and 62 per cent of the employment in

Table 5.4 Anticipated growth of the four metropolitan areas, 1980–2010 (millions)

	1980	2010
PWV	6.07	16.23
Durban	1.96	8.04
Cape Town	1.79	4.59
Port Elizabeth	0.69	1.76

Source: Simkins (1990: 224)

the tertiary sector is located there. Within the PWV, the Central Witwatersrand's position is indicated by its providing about half of the employment opportunities. In total, the PWV contributes 41 per cent of the country's economic output. If considered in conjunction with its adjacent regions, the South-Eastern Transvaal coalfields and the Northern Orange Free State goldfields, the broader region contributed 60 per cent of South Africa's economic activity. When presented in terms of resident, migrant and commuter labourers (in the latter two cases, a dependency ratio of 3:5 is assumed), it appears that the broader region is directly supporting 40 per cent of the country's population (Naude 1986).

These figures are impressive, but they are best viewed in comparative terms. On the basis of international comparison, South Africa is seen to benefit from a marked lack of domination by any one centre (Richardson 1977: Table 2; Gilbert 1981). In Renaud's (1979) comparison of the 'degree of urban concentration' of 125 countries, South Africa had the ninth lowest degree of concentration. These comparisons are a bit misleading because they disaggregate the cities comprising the PWV and consequently understate the PWV's dominance, but, this notwithstanding, South Africa's economy and population cannot be viewed as excessively centralized.

It is difficult to rationalize regional policy, and this is all the more so because the study of urban systems around the world suggests a distinct pattern of urban development as nations develop. The pattern is described by population movement, which reflects the relative rate of economic growth of different cities and regions. Initially people move inward from periphery to core and from smaller centres to larger ones. Thereafter there occurs a series of 'polarization reversals'. People first move from the central city to the now more rapidly growing surrounding urban tracts. At much the same time, the growth of the primate city slows and secondary cities enjoy a higher growth rate. Next, as in Western Europe and the United States, the large cities throughout the urban system begin to lose population. Last, in a move Hall (1987: 239) titles 'the Clean Break with the Past', cities generally lose population and there is a reversal of rural–urban migration.

The implication is that regional policy contributes less to regional development than does a high rate of economic growth. This position has a venerable lineage. A compilation of theories postulates an inverted U-shaped curve, where income inequality (suggested by Kuznets during his 1950 presidential address to the American Economic Association),[7] regional income inequality (Williamson 1965) and urban primacy (El-Shakhs 1972) first increase, then decrease, with development. As noted, the significance of urban primacy follows from the belief that a non-primate or dispersed system of cities helps 'transmit' growth to a nation's periphery. Despite this lineage, the theories have uncertain validity. For example, a decline in urban primacy is common among most, but not all, developed countries, and does not appear, on its own, to be a precondition of a reduction in regional differences in income and employment levels. In general, while interpersonal and regional income inequality do appear to have increased and then declined in the developed countries, it is unclear that the same will necessarily prove true in less developed countries. Gilbert (1982b: 177) suggests that 'convergence is likely to be weak because: today's poor countries may never reach the levels of per capita income at which regional convergence begins; regional disparities in less developed countries are much greater than those characteristic of developed countries in the past; and convergence depends on effective government intervention and many governments show little sign of interest or ability to remedy regional inequalities'. The quote is interesting because Gilbert proposes that there are automatic processes if a country reaches an (unspecified) level of per capita income and if a country has a government that is concerned to address regional inequality.

Where does this leave South Africa? First, to the extent that a non-primate urban system is the vehicle for overcoming regional inequality, South Africa is blessed with just such a system. Secondly, given that South Africa is a middle-income country that is concerned about regional differences, it seems clear that both Gilbert's conditions are realized. The application of the U-curve theories to South Africa causes one to doubt the need for a decentralization policy, as opposed to an equitable allocation of resources. Both sides of the coin give the same indication. Where the PWV is concerned, it is almost certainly the case that, as 'in today's largest cities . . . urban economies exceed urban diseconomies' (Gilbert 1982b: 177). The PWV is better endowed than the rest of the country with the management skills and economic resources necessary to address poverty. In effect, this area is where jobs will be most efficiently provided. Informal sector opportunities, too, are heightened in large urban concentrations like the PWV. One's confidence in holding these positions also has to do with its scale. In the case of larger urban agglomerations, one has sympathy for equivocations regarding the benefits of unhindered growth, but the PWV, in international comparative terms, consists of a number of relatively small cities. Moreover, the total increase in the country's black population between 1980 and 2000 is equivalent to less than the current size of the population of a number of cities around the world. The

PWV is not facing problems related to scale as much as it is facing the difficulty of overcoming the distinctly inefficient, and expensive, form of the apartheid city.

Alternatively, in respect of the South African periphery, Bell's data suggest that market-induced processes are already in place that are integrating distant centres into the economic mainstream. In terms of the above findings, it would appear that the more rapid the country's rate of economic growth, the more rapid will progress be towards the lessening of regional inequalities. The fact that South Africa already has a well-developed urban hierarchy lends further support to the position that industrial growth will benefit the regions. There are not the inhibitions, the lack of urban facilities and the absence of a local market that are so common in Third World countries.

A last argument, concerning likely trends towards reduced inequality, follows on neoclassical assumptions regarding labour mobility. A trend towards the greater spatial equalization of employment and income levels is now likely since, with the progressive removal of mobility controls on blacks, they will be able to escape their capture in regional pockets of poverty. In other words, poverty-stricken regions, such as the Eastern Cape, need the PWV's growth in order to spur migration and to reduce unemployment and the demand for services.

Recommendations

The current decentralization programme should be abandoned and, indeed, it appears that it is in the process of substantial revision. With a view to the future, two premises should underlie regional policy. One is that one cannot, and should not, attempt to control or divert growth from the PWV. Quite the contrary, its growth should be encouraged. The other is that, in a democratic South Africa, there will still be calls for policies that are intended to promote development in the periphery. The goal, here, is to create advantages for the periphery while not damaging the economy of the core.

Growth pressures on the larger cities are inexorable. In order to cope with these pressures, South Africa needs an urbanization strategy that includes a variety of policies pertaining, for example, to housing, services, the local economy and the expansion of the municipal tax base. In so far as the location of industry is a relevant concern for the larger cities, this is an issue that should be decided by local and metropolitan government. Ideally, the location of industrial zones and investment in infrastructure will be directed at re-concentrating the cities and at overcoming the legacy of the apartheid city. Industrial deconcentration may well become a relevant planning goal in the future, as the cities grow, but this is a local or metropolitan issue and does not concern a national urbanization strategy.

There is no reason for persisting with attempts to direct industrial investment to South Africa's peripheral areas. While one can expect that there will be pressure to steer investment to the homelands, they lack large centres and centres with industrial potential. They are, correctly, a source of migrants. Moreover, the homelands generally abut an adjacent city with potential, such as Durban, East London or Pietersburg, and blacks living on the edge of these cities already benefit from their growth. Spatial chauvinism should not lead to less efficient sites for investment. Growth in the homelands is desirable, but not at the expense of economic efficiency. Moreover, when the Group Areas Act is withdrawn, the rationale for decentralization will diminish further. Elsewhere, in homelands like Venda or the Transkei, where there is no pre-existing reason for industrial development, migration should be accepted as an inevitable process accompanying development. The primary issue here is not the location but the growth of employment, accompanied by the freedom to live where one chooses. Interference in the process of growth, especially the kind of interference that takes jobs to low-wage regions and reduces the black share of the national income, is impossible to justify.

A prominent means of targeting development to areas containing many impoverished people, without, at the same time, acting against the core, appears to be a secondary city strategy. This strategy presumes that urban economic development in such cities has been retarded in the past, anticipates rapid urban growth, and seeks to create alternative migrant destinations. The type of centres suggested are Port Elizabeth, Bloemfontein, Pietermaritzburg, Pietersburg and East London. There should be two components to government policy. One would ensure the finances for locally initiated projects aimed at improving the social and economic infrastructure of the cities. For example, there should be investment in land and housing that is based on the small-scale construction industry and that fosters growth in the household durables market. The second component would consist of the government trying to create an environment that enables local enterprise to increase its efficiency of operation and to improve its competitive posture in national and international markets. A policy such as this might slightly diminish regional differences and even perhaps influence migration flows, but the effects are likely to be marginal. Regional planning has always offered more than it can deliver.

Notes

1 F. W. de Klerk, the new leader of the NP (cited in *Business Day*, 9 February 1989).

2 A primate city is one that is markedly larger than the nation's other cities.

3 The concessions are of two sorts: long-term concessions intended to compensate for the inherent disadvantages of the location; and short-term incentives designed to alleviate financing problems during the initial years at the location. A comparison of some of the concessions between Brits, a deconcentration point outside the PWV and Dimbaza, a decentralization point in the Ciskei is as follows.

	Brits	Dimbaza
Long term:		
Rail rebate	no	60%
Housing subsidy	20%	60%
(of interest rate)		
Short term:		
Wage subsidy per month	R30	R110
(Paid quarterly, cash, tax free)		
Rental subsidy of premises	20%	80%

4 For example, in 1985 wages for semi-skilled labour in decentralized locations in Natal were equal to 69 per cent of those in non-decentralized locations and wages for unskilled labour in decentralized locations were 58 per cent of those in non-decentralized locations (Addleson *et al.* 1989).

5 This was a point frequently emphasized at the workshop 'Industrial trends and prospects in Region E', held at the Elangeni Hotel, Durban, 11 November 1987.

6 The cost was estimated in the following manner. The Development Bank of Southern Africa's 'latest figures' are not always for the same year; none the less they show 43,448 manufacturing jobs in the independent homelands and 46,477 jobs in the other homelands. It seems reasonable therefore to assume that expenditure on the incentives in the two sets of homelands is likewise about equal. (Some investment in the Ciskei is occurring in terms of the free enterprise zone scheme being implemented there, but the bulk of the investment there has occurred under the incentive scheme.) Approved applications, jobs and investment in the non-independent

homelands during the first five years of the programme were respectively equal to 24 per cent, 44.4 per cent and 29 per cent of all totals reported by the Decentralization Board. I am assuming that the figures in the independent homelands are more or less the same, but this may well be an underestimate because the labour concessions are higher in the homelands and because they are more out of the way and presumably require greater concessions in respect of transport. My estimate of total expenditure on incentives is obtained by multiplying R505 million by 1,444, which gives a value of R729 million.

7 Although recently Ram (1988) has criticized the Kuznets hypothesis.

6 Urban and rural development in the periphery

Rural development strategies intended to improve the incomes available from farming are often viewed as an integral part of an urbanization strategy since it is felt that they will restrain rural–urban migration. Yet there 'is no evidence, either in Africa or elsewhere, to back up this belief. Rural development strategies are critically important in developing countries, especially in Africa, but this is not because they offer a solution to the problems of urbanisation' (Brennan and Richardson 1986: 35). Rural development, with the exception of only a few types of agricultural production, is labour displacing. It is, after all, a defining characteristic of a country's economic advance that a smaller proportion of its labour force is located in rural areas, engaged in agricultural pursuits. However, in a situation where there is extreme inequality in the ownership of land, the redistribution of farms may make it possible to settle a larger population on the land. In the light of international experience in less developed and middle-income countries that small farms can have higher levels of output and employment per hectare than large farms, it is extremely interesting to ask whether the same would be true in South Africa. Should one advocate the redistribution of land?

Much of this chapter is taken up with arguments surrounding this issue. They reveal that, while there are excellent reasons for redistribution (for example, increasing black capital accumulation and decreasing hunger), they are flawed in the case of urbanization policy. Redistribution should not be viewed as a strategy that will both improve agricultural productivity and also increase the density of the population on land presently owned by white farmers. Instead of redistribution, I suggest government legislation in respect of wage levels and working conditions, but neither will restrain urbanization pressures. Legislation that improves the conditions of labour serves to displace those labourers whose marginal product is below the social wage and to increase the drift to the towns.

In the second half of the chapter I consider whether economic development and urban growth in the homelands can check urban migration from them. A study of this sort will be most exacting in a homeland that is distant from large urban centres in the rest of South Africa and that has a good agricultural base, for here the potential for agricultural development is clearly greatest. The Transkei meets these criteria and it is employed for illustrative purposes, but I show that even in this instance agriculture will not take up more than a third of the labour force. The critical question in the homelands is whether there will be substantial movement to the cities or whether many will choose to remain in urban centres and rural settlements in the homeland. Since it seems that a substantial number of people are, in fact, urbanizing within the homelands, the last part of the chapter addresses how urban development in the Transkei can best meet the needs of the poor.

Land redistribution

There are many forms of land redistribution and I have considered only the form that is likely to increase the population density on the land most – thus, the following material is not intended to reflect on the desirability of land redistribution per se. While one should note that the Freedom Charter and the African National Congress's Constitutional Guidelines do, in an ambiguous fashion, call for redistribution, the issue at hand is whether the redistribution of large farms to small farmers will restrain urban migration. However, the viability of land redistribution of this sort cannot be considered independently of a concern for the impact of redistribution on national agricultural performance. Accordingly, I reflect on arguments for and against redistribution to smallholders and make judgements in this respect, but try none the less to lead the arguments into an exploration of the implications for urbanization policy.

There were only 58,500 white farming units with a total investment per unit of R762,000 in 1986. This represented a decline of 34,400 units between 1967 and 1986 (*Star*, 4 May 1988: 11). As the land under cultivation declined by only 5 per cent during the period 1970–1985, some of which was probably due to the transfer of land to the homelands, the large reduction in the number of farm units was due to the consolidation of farms. The area farmed is about 85 million hectares and the average farm size is 1,438 hectares (M. Cobbett 1986).

Data regarding land holdings in the homelands are shown in Table 6.1. There are 2,338 million hectares of arable land in the

Table 6.1   The size of homeland land holdings

Homeland	Surface area (000 ha)	Arable land (000 ha)	Per cent arable	Pop. (000)	Arable land per h/hold (ha)
Bophuthatswana	4,000	400	10	1,721	1.4
Ciskei	650	75	12	750	0.6
Gazankulu	773	65	8	620	0.6
Kangwane	300	36	12	448	0.5
KwaNdebele	197	24	12	286	0.5
KwaZulu	3,316	565	17	4,382	0.8
Lebowa	2,454	347	14	2,157	1.0
Qwaqwa	62	7	12	209	0.2
Transkei	4,355	754	17	3,000	1.5
Venda	620	65	11	460	0.8

Source: M. Cobbett (1986: Table 1).

homelands, 14,033 million people (2.34 million households) and a weighted average of 1 arable hectare per household. (The surface area of South Africa as a whole is 2.9 hectares per capita, but much of this is desert or semi-desert.) One way of interpreting the table is to consider the land needed to sustain a family. The staple diet in the homelands is maize and one family requires about 1,000 kg of maize a year. With current productivity levels in the homelands of about 500 kg per hectare (350 kg in the Ciskei), it takes 2 hectares to produce this maize. As a matter of interest, the average output on projects in the homelands is about 1,900 kg per arable hectare.[1]

Even if one accepts a 40 per cent urbanization level for the homelands (Graaff 1986), and the fact that most of the 14 million blacks in the homelands in 1985 represent a form of displaced urbanization and do not seriously undertake agricultural pursuits, one gains a sense of the distributional inequality. In larger homelands, like KwaZulu, landlessness may equal a third to a half of all households while in cases like Qwa Qwa, where, due to relocation, the population growth between 1970 and 1982 was nearly 2,000 per cent, the situation is much more extreme (South African Institute of Race Relations 1985).

OPPORTUNITIES FOR, AND OBSTACLES TO, THE REDISTRIBUTION OF LAND
Some reports say that there are a number of 'abandoned' or 'vacant' white farms in peripheral white farming areas and, if this is the case, there are ready opportunities for redistribution. For example, Greenberg (1980) mentions that, as early as 1960, 24 per cent of the

farms in the Orange Free State were 'unoccupied' and that in four northern Natal districts the figure was 42 per cent. These dated figures are unlikely to have lost relevance because the area being farmed by whites has contracted since the 1960s, but one should question whether they are accurate. For example, the land may have a black manager, with little or no commercial farming being undertaken, or it may be occupied by squatters – in both instances the land is occupied. Such land is obviously a prime candidate for redistribution but, to the extent that the land is already occupied by blacks, its subdivision may result in only a transfer of title, rather than the settlement of additional population. Moreover, given its geographical and physical characteristics – arid, with poor access to markets and infrastructure – this is equally land that anyone would find difficult to develop for commercial purposes. Nevertheless, the reallocation of abandoned land provides an immediate and cheap opportunity for reducing the inequality in land holdings and, perhaps, also for relieving population pressure in the homelands.

I discussed this position with two professors of agriculture at the University of Fort Hare in the Eastern Cape,[2] where there have also been reports of abandoned farms (*Surplus People's Project*, n.d.). They flatly denied the existence of such land; instead they noted that, while the white owners might not live on the land, it would be occupied and managed by a black foreman. Alternatively, they held that a farm might have been attached by a financial institution to whom the farmer had become indebted. In other words, they could understand why people might think that a farm was abandoned, but insisted that this would be mistaken. The one exception arises when squatters might occasionally occupy farms that have been acquired by the South African Development Trust (of the Department of Development Aid) with a view to handing over to the relevant homeland. Another rural expert in the Eastern Cape confirmed squatting on formerly white farms prior to their acquisition by the homeland government.[3]

It appears that it would be incorrect to assume that a cost-free alternative exists. Except in the Northern Transvaal, where farms may be unoccupied owing to the insurgency, the existence of abandoned farms should be held in doubt.

What are the obstacles to the redistribution of land? (It might be better to say the *further* redistribution of land.) On the one hand, concern with the size of black land holdings has been a perennial issue and the 1913 Land Act incorporated measures intended to reduce the tenancy options on white farms and to impose a shortage of land on blacks so that they would seek wage labour on white farms as well as in the mines and the towns. On the other hand, it was recognized that excessive land deprivation would promote the urban drift of the homeland population and, since the 1936 Development Trust and Land Act, 5,434,109 hectares have been acquired from whites with a view to geographically consolidating and increasing the size of the homelands (South African Institute of Race Relations 1987). In a sense, the 1936 Act has already prepared the ground regarding the political feasibility of land redistribution; one might say that there is already a legal and institutional precedent for the redistribution of land in South Africa.

Furthermore, in respect of their voting power, 58,500 white farm units do not add up to many votes and, since farmers largely support the National Party and the more right-wing Conservative Party, they are hardly likely to constitute a relevant constituency for a democratic government. If the power of farmers is not political, it must then be sought in their economic role. Although the drift of much of the following discussion may at first seem to indicate that large farmers can effectively be replaced by small farmers and that the position of the former is rather weak, my argument, as it progresses, leads to the contrary conclusion – that large farmers make an important contribution to the economy that is unlikely to be replaced by small farmers. This is a highly contentious conclusion; for example, one can find both sides quoting Zimbabwean data in contradiction of each other. My enquiry, reflected below, has led me to the view that, were small farmers to replace large farmers, the likely outcome would be a loss of agricultural self-sufficiency, higher food prices and a foreign exchange burden due to agricultural imports. Because of their economic significance, the government might well be loathe to redistribute the land of the commercial farmers.

However, to dispute the ease of redistributing land is not to claim that there is no opportunity for doing so. For instance, in 1985, 0.9 per cent of the farming units produced 15.9 per cent of the gross farm income, 5.8 per cent produced 38 per cent of that income and 27.5 per cent produced 73.8 per cent (Hattingh 1986). These data suggest that, were the government to consider redistributing land, it could do this on the more marginal farming units without disrupting the bulk of the country's agricultural production. Hattingh, however, notes that the relative contribution to gross farm incomes is more or less paralleled by the distribution of land ownership.[4] For instance, at the bottom end, 30 per cent of the farming units occupy only 3.5 per cent of the land. Thus, while some land redistribution can take place without affecting the farming units that produce the bulk of the country's agricultural output, if the redistribution occurs on a large scale then the government would soon have to confront major decisions regarding the goals, costs and organization of agricultural production.

My hypothesis is that the stumbling block to the redistribution of land by an ANC-led government would have less to do with directly political opposition from large farmers than with their economic role and, importantly, the low political profile of the rural poor. In other words, I am suggesting that one should not expect that the government's own prioritization of resource allocation would favour the rural poor. Certainly, in an industrial and increasingly urbanized economy, there does not seem to be much precedent that would lead one to hypothesize that the redistribution of land would be a prominent policy. (This is not to deny that the government would like to see a racial change in land ownership, but this is not the same as a redistribution of land.)

As a result of these misgivings, the first part of this chapter is largely hypothetical. The issue being addressed is whether it is possible to enable rural households to obtain an income from the land that is sufficient to prevent them from migrating and that does not involve a decline in output per hectare. Might it be that, for once, the goals of employment creation, equity, efficiency, regional and rural development and restrained urbanization may be compatible?

The argument for land redistribution

This section contains arguments for redistribution, including repeated comparison with South African circumstances, and the next section continues with my doubts and uncertainties.

The International Labour Organization (1972: 166) (ILO) has shown in Kenya that output per hectare declines with increasing farm size and suggests that there is 'a strong presumption for thinking that both employment and output tend to be higher on smaller farms'. Similarly, Hunt (1984: 230) argues that 'land reform which creates more equal individual access to land can increase both rural equity and the social efficiency of agricultural production. This is a widely held view, and most of the arguments for it have been well rehearsed'; and, in a comparison of productivity on small farms, the World Bank weighs in with the view that 'The smallest-scale farmers are the poorest. Paradoxically, they are also the most productive because their cultivation is most labor-intensive. . . Efficiency would improve – and poverty diminish – if small farmers bought or leased land from large farmers' (*Research News*, 1986, 6, 4).

These conclusions are not universal – in India, there is a loss of efficiency at the very smallest farm size. This one variation in the data notwithstanding, similar results were achieved following the land redistribution schemes implemented in Japan, Korea and

Taiwan (all after the Second World War) as well as in Mexico, Bolivia, Chile and Venezuela. The only failure recounted by Berry and Cline (1979) occurred in Peru, where the redistribution took the form of state cooperatives and involved poor incentives for members of the cooperative. (They do, however, report many failed schemes because bold ambitions were not followed through. It seems that slow, incremental implementation endangers redistribution schemes.)

SCALE ECONOMIES AND THE RELATIVE COST OF LAND, LABOUR AND CAPITAL

An initial equivocation in respect of these findings would be that while output per hectare might increase with smaller farm size, costs per hectare would increase more than proportionately. One would ordinarily anticipate scale economies, which means declining unit costs of production as the scale of production increases. However, Berry and Cline (1979: 5) assert that repeated empirical enquiry reveals that the expected returns to scale are 'approximately constant in developing country agriculture and, therefore, [are] neutral with respect to the more general issue of farm size as related to productivity'.

Many would expect that scale economies would also be achieved as a result of the mechanization of large farms and their supposedly better management and willingness to adopt new innovations. Instead, it is argued below that imperfections in capital markets and the underpricing of capital, relative to land and labour, have had much to do with the prior, and excessive, mechanization on large farms in South Africa relative to the costs of machinery, land and labour. Further, while large farmers 'are likely to be the first to adopt innovations, small farmers are likely to follow – and sometimes to do so very soon' (Berry and Cline 1979: 28). Once the adoption of new techniques has taken place, 'the earlier size–productivity relationship is likely to be re-established in relative terms, with a higher output per [small] farm area at all sizes' (idem.).

There are a number of contrary reasons for the finding that small farms may demonstrate superior efficiency in the use of land, labour and capital. In large part these centre on the fact that the 'effective price' of capital and land is low for the large farms and high for the small farms, while the reverse is true for labour. The point has very specific implications for the intensity with which land, labour and capital are employed.

In an agricultural structure composed of very large estates holding most of the land on the one hand, and a large number of small farms on the other, agricultural production tends to be below its maximum potential level because land is underused on the large farms, while excess labor without fully productive work is crowded onto the small farm sector (or in the landless labor force). Agrarian reform redistributing land from the large estates into new family farms of moderate size can combine the underused land with underused labour and raise production (as well as total capital and seeds–fertilizer requirements).   (Berry and Cline 1979: 7)

These findings are not duplicated in South Africa: there are apparently no data on variations in land utilization on large farms and land in the homelands is not used intensively, although there are very specific reasons for the latter which we will come to shortly.

In the case of capital, it is a widespread and well-known phenomenon, both in the First and Third World, that sources of capital prefer larger borrowers, who are perceived to create less risk and who also entail smaller transaction costs relative to the size of the loan. The same is doubly true where minorities or women are concerned and, in South Africa during the entire period prior to liberation, for blacks. Besides, the superior access of large farmers to capital, government credit and subsidy programmes usually artificially lowers the cost of capital, relative to land and labour, and may lead to the premature importation of equipment. The result is a misallocation of resources, with large farmers squandering capital and foreign exchange. This misallocation of resources has been particularly pronounced in South Africa where white farmers currently support a debt of about R14 billion (although apparently R5 billion of this is household, as opposed to farming, debt).

Lastly, with regard to labour, many would probably accept the observation that mechanization on large farms will often be undertaken in advance of relative price changes as a result of the attitudes of farmers regarding labour relations and also because of subsidized mechanization which the government introduced with a view to reducing the number of blacks in white South Africa. Here is one explanation of the low labour intensities frequently found on large farms.

If one adopts more of the stance of textbook economics, a large farmer will employ labour until the value of the output of the last labourer employed is equal to the prevailing wage. It would be irrational to employ more labourers, for then the farmer would be paying the labourer more than he or she produces. In the case of the small farm, however, the peasant will continue to work even where the value of his output is lower than the wage paid to labourers on large farms. Where the family's dominant objective is to produce enough in order to survive, the peasant has every reason to work beyond the point where labourers cease to be hired on large farms.

The objection to this is: why does the peasant not realize, at the point where the value of his or her labour is worth less than the large-farm wage, that he or she would gain greater returns by ceasing to work on the small farm and by seeking a job on the large farm instead? One reason is that peasant families share their output so that, while the value of the individual peasant's output may be low, in order to forgo his share of the family output he would want a wage that is above the level of his own output and that is closer to the average output of labour on small farms. Other considerations deal with living costs off the peasant's farm and transport costs to the large-farm job. In addition, the probability of actually finding a job may be low, so that the family member's calculation is based on the wage relative to the probability of getting the job and, with minimum wage legislation that raises both wages and the farmer's reluctance to hire additional labour, this probability may be quite low. Moreover, jobs on large farms tend to be seasonal, and occur at precisely the times when labour on small farms is most in demand. One can surround the point with numerous equivocations, but it is a correct conclusion that it is almost invariably the case that labour inputs are higher on small than on large farms (Mazumdar 1983).

In the homelands the situation is different because the labour force is oriented to selling its labour outside the homelands rather than working 'on the farm'. A post-apartheid environment will differ in at least three respects: unemployment will unavoidably have worsened; redistributed smallholdings might be sufficiently large to ensure a reasonable family income; and farmers will have access to the rural support services and infrastructure that necessarily should accompany a land redistribution scheme.

It might be that, even acknowledging the above conclusions, a sceptic could maintain that they do not apply to South Africa because they are relevant to the Third World, whereas South Africa is popularly held to combine the First and Third World, with white agriculture being characterized as typical of the First World. This objection is an important point since the relationship between farm sizes ceases to apply once a country reaches a certain level of development.

The point at which the superior position of small farms ceases to apply occurs when there is a sufficient number of reasonably paid off-farm employment opportunities resulting in less pressure to use labour intensively on the small farm. If this is coupled with better access to credit, extension services and markets, then the differences between large and small farms will further diminish. The question in South Africa is whether this levelling-off has been reached and, even if it has, whether the economy is not deteriorating, so causing it to enter a phase where small farmers might be expected to have superior productivity. For most of this century, smallholder agriculture has been in an extremely disadvantaged position, but the lack of agricultural opportunities was ameliorated by alternative employment opportunities. The post-apartheid environment could be different since, when considerably greater unemployment is coupled with improved smallholder opportunities, it may be that many might want to farm.

FARM SIZE

Is the redistribution of farm land feasible? In particular, how much land is necessary to create farming opportunities and how much land is available? Berry and Cline (1979) suggest dividing the total farm land available by the total number of families in the rural labour force. However, this is unlikely to be an acceptable method in middle-income countries since the result might well provide an income level lower than the urban alternative (in relation to the probability of obtaining such a job). For instance, if one follows Berry and Cline's suggestion in South Africa, an arable land measure of about 0.5 hectares per household, or about 1.0 hectares if we assume a 50 per cent urbanization rate, will not enable households to earn an agricultural income that is large enough to attract them to the land.

Hunt (1984) suggests that one should first determine the farm size necessary to provide an income equivalent to a poverty-level income and then divides the farms accordingly. Her very low targeted income is understandable, given to the Kenyan context, but the South Africa case is different because of the superior urban opportunities. The situation is one of a dynamic interrelationship between farm size and incomes and urban opportunities. As the one set of incomes and expectations changes so, too, the other will change. The adjustment one should expect, over time, is the accumulation of farm land in fewer hands – an inevitable outcome if urban incomes rise and a proportion of the rural population leaves for the cities.

The following figures provide an initial idea regarding the appropriate size of a small farm, but they (and the figures that succeed them) should be taken as pointing to an order of magnitude alone, since soil and micro-climate variations would make a mockery of claims of precision. Bembridge has indicated[5] that in the Eastern

Cape, for example, about 4 hectares of arable land and 24 hectares of pastoral land should be sufficient for a household for subsistence purposes. Apparently a reasonable presumption is that 12 per cent of the Eastern Cape is arable. The amount of land necessary to restrain migration would, accordingly, be some combination of arable and pastoral land and each smallholding would be a multiple of the subsistence minima just indicated. How much land is there?

The magisterial districts of Queenstown, Cathcart, Stutterheim and Komga, shown in Map 6.1, provide a good example as they are located in the 'white corridor' between the Transkei and the Ciskei and are likely to bear the brunt of the future demand for agricultural land. The average farm size, number of black workers,

Map 6.1 The Transkei and the 'white corridor'

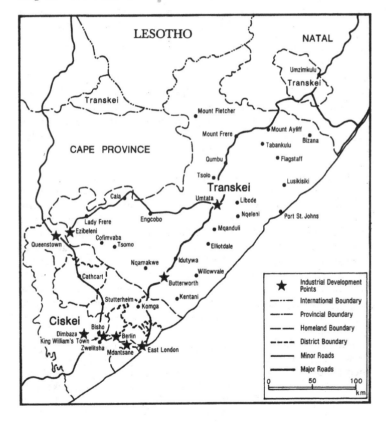

land per black worker, and value of the land are shown in Table 6.2. The amount of land per employee ranges from 99 to 51 hectares. The districts with smaller areas per employee are more arable. The value of the farm divided by the number of employees varies between R11,868 and R19,583 per employee.

These statistics create a false impression since there is often more than one employee per family on the farm. In a micro-study of farms in the close-by Albany and Bathurst districts, Manona (1985) found that the situation varies: on some farms 'virtually all able-bodied men and women living there can always find employment throughout the year' (p. 6); but elsewhere that, 'Except during the shearing season, there are no jobs for women on the farm, apart from domestic work' (p. 8). Regardless of the number of paid farm workers per family, it is obvious that the correct equation is based on the number of households, not on the number of employees. In this respect, there are currently 9,223 black employees in the four districts, of whom 1,087 are domestic employees and 3,532 are casual employees. In other words, the number of households in the districts is probably closer to 6,778, or about 40,668 persons when multiplied by six.

It is apparently a common mistake to base the redistribution of farm land on its division among existing farm households. While

Table 6.2 Land available per worker in the white corridor[1]

Magisterial district	Average farm size (ha)	Regular employees	Average no. of casual employees	Domestic employees
Cathcart	1,786	1,258	1,056	289
Komga	773	941	777	223
Queenstown	1,255	1,142	690	298
Stutterheim	671	1,263	1,009	277

Magisterial district	Average value of farm (R)	Value/no. of employees (R)	Area per employee (ha)
Cathcart	324,927	16,840	97
Komga	190,168	15,382	63
Queenstown	248,295	19,583	99
Stutterheim	155,936	11,868	51

[1] The table includes only black employees. The value of the farm is that imputed by the owner.
Source: South Africa (1981b).

farm labourers are welcome, and indeed would be encouraged, to acquire the land they work, this is not the issue here. The question is whether there is sufficient land to 'absorb' population in addition to those people already on the land. The ensuing calculations suggest that the answer is in the negative, which leads me to the arguments against redistribution.

The argument against the redistribution of land

As noted earlier, the issue is not simply one of enticing people to remain in the rural areas, it is also to enquire whether they are likely to use the land productively. South Africa is a food-surplus and food-exporting nation – if balance of payment and debt problems are not to be exacerbated, there is every reason to wish to remain so.

IS THERE AN ABILITY AND WILLINGNESS TO FARM?
The uncertainty that arises concerns whether current productivity levels reflect on the future. There are many reasons for the present low output levels in the homelands. For instance, agriculture among blacks has suffered from such deep disadvantages that in the Transkei, where 91 per cent of the population lives outside proclaimed towns and in which about 20–30 per cent of the population is landless (Baskin 1984), 40 per cent of the arable land is not farmed.[6] Moreover, Bembridge's[7] survey of three Transkeian districts revealed that about 50 per cent of the land that was ploughed was farmed only in order to obviate moves by the government to appropriate under-utilized land![8] This unwillingness to farm has many explanations: the unequal distribution of use rights to land, with the holdings of many being uneconomical; better returns to migrant labour; the absence of a marketing network or credit facilities; the general lack of draught animals[9] and an ability to afford tractor services; insufficient family labour; the many competing demands on women's labour; unwillingness to grow crops when this might result in a reduction in the level of migrant remittances; poverty and an inability to afford implements and seeds . . . and so on.

It appears that perhaps up to 20 per cent of the rural population wish to farm, while the balance wish to earn a living through alternative means.[10] Thus M. Cobbett (1986: 18) has held that the 'land problem is . . . more of a perceived rather than a real or objective problem'. He argues this because of 'the structural dependence on urban derived earnings and the relative unimportance of earnings from agricultural sources'. For example, commercial and subsistence agricultural production represented only 6 per cent of migrant and commuter earnings in 1985 (calculated from M. Cobbett 1986: Table 3). In addition, in his view, a location in the homelands does not denote agricultural skills, nor does it denote a desire to farm. These are tremendously important points because, if rural blacks lack the ability and inclination to farm, then the earlier Kenyan and other comparative material are of little relevance. Cobbett, however, was sharply criticized at the York conference on a post-apartheid South Africa.[11] Is it that the critics were largely white and leftist and that their views were distant from those of the people? Or is it that only 20 per cent indicate a desire to farm because of the limited nature of current opportunities? If land, credit and support services were available and if there was access to markets, would more people want to, and be able to, farm?

Surveys in the Transkei, the Ciskei and KwaZulu further emphasized uncertainty regarding the willingness and ability of rural blacks to farm. Pertinent conclusions were that 'The type of small farmers that are found in the rest of Africa and other developing countries were not identified' (*Ardrinews*, 7, 2: 1); and that 'It would seem . . . that a broad agricultural strategy aimed at the rural population as a whole has little chance of success, however vigorously it may be promoted, because the bulk of the population do not, and would not like to strive to generate income from an agricultural source because they prefer to fulfil their income needs from non-agricultural sources' (ibid.: 2). This use of rural land as a place to live, rather than as a productive resource, was evident in the Ciskei, admittedly a bad case of land deprivation, where, in three districts, the sources of income were: non-agricultural – 94.4 per cent; subsistence production – 3.8 per cent; and agricultural income – 1.8 per cent (*Ardrinews* 1987, 7, 2: 1).

When wage employment (typically migrant and commuter labour) and pensions are almost the exclusive source of income, it would be surprising if people retained the agricultural orientation and expertise that they demonstrated in the nineteenth century. The priority list of people in the Ciskei, Transkei and KwaZulu ranked items in the following order: domestic water, roads/transport, employment, improved housing, clinics/creche, sundry agricultural needs, pensions and more schools. Thus, while from an agricultural perspective there is tremendous overcrowding and insufficient land to earn a living from farming, it seems that this is not the overriding perspective of the people themselves. The issue is not one of farming opportunities but one of improving roads and access to water, clinics and housing.

At the very least, this uncertainty argues that the wholesale redistribution of land should be avoided, as there may well be a lack of willing farmers. None the less, if only 10 per cent of the black population wishes to farm, this is still a sizeable number. Should redistribution in favour of the 10 per cent be facilitated? There are two aspects to the answer: one dwells on the possibility that, given more favourable conditions, small farmers could demonstrate the productivity levels of large farmers; the other rests on the ability of the land in the white corridor to sustain an increased density in population.

FARM SIZE AND PRODUCTIVITY CONSTRAINTS
In the former regard, an interesting comparison comes from Zimbabwe where the postwar contribution by many peasants to the output of cash crops has been heralded. Output levels of maize, reported in 1985, in Zimbabwe were of the order of:

- 1.5 tons/hectare in communal areas (average land holding 4.5 hectares);
- 2.1 tons/hectare in resettlement areas with good soil and access to services (average land holding 20–60 hectares, of which 6 hectares are arable); and
- 5.7 tons/hectare in commercial farming areas (average land holding 2,200 hectares).[12]

Maize is the crop that could be expected to provide the most favourable comparison since it is a familiar food crop and does not presume differentiated technology or large capital inputs. Even so, large farmers have more than double the productivity level of small farmers, and the Zimbabwean government would risk the health of the economy were it to undertake a more aggressive programme of redistribution.

The trouble with these figures, according to Weiner *et al.* (1985), is that large farms are located on high-potential land where there is agricultural infrastructure, whereas the reverse is generally true for small farmers. When they compare large and small farmers on land of equivalent potential they find that the output of maize for small and large farmers is comparable. In conversation, though, Weiner continues that the same is not true for certain other crops, in particular virginia tobacco, which benefits from sophisticated managerial and capital inputs. Does this leave one with an argument for the redistribution of maize farming land?

It is difficult to answer in the affirmative. For example, Hayward provides less favourable data.[13] He found in Zimbabwe, on land of equivalent quality that, while the top 15 per cent of small farmers had levels of output equivalent to the large-farmer average, the average for all small farmers was 50–60 per cent that of large farmers. Further, when weather conditions varied from the norm, this figure declined. The competing views suggest that Weiner *et al.*'s position is best thought of as exploratory and not, in itself, an argument for redistribution.

The situation in South Africa is further complicated by the fact that most blacks who are located in rural areas do not want to farm and it is unclear that those who do can farm efficiently. With a few isolated exceptions there is not a willing peasantry waiting in the wings. It might be that with increasing unemployment and better farming conditions many more might want to farm – but this is a hypothesis best explored through equal access to land and a small-farmer support programme than through wide-scale redistribution.

In general, Hattingh urges against a singular focus on farm size and productivity.[14] He argues that in South Africa it would be incorrect to assess optimum farm size through an exclusive focus on output. The goal of maximizing agricultural output may be relevant in a country suffering from food shortages but, in a country having food surpluses, the more appropriate goal is maximizing agricultural income. Figure 6.1 combines the three essential measures for commercial farms: output per hectare, cost per hectare and farm size. While output per hectare may be higher with smaller commercial farms, so too are the costs. In other words, returns to scale are not constant The optimum farm size for the country is the one that maximizes the distance between the two curves, since a redistribution of land that would reduce the distance would

Figure 6.1 Ascertaining optimum farm size

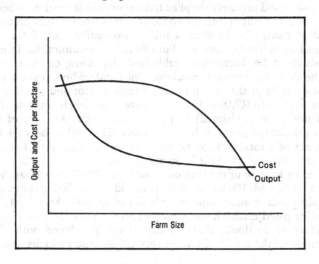

concurrently reduce the income earned from the land – with negative consequences for the national economy.

IS THERE SUFFICIENT LAND?
In the light of these considerations, is there sufficient land in the white corridor to settle more people than are presently on the land? In order to attract people to the land they must be able to earn an income that is competitive with urban incomes, which means that the farmers will have largely to be engaged in commercial farming. Commercial farming obviously requires planting on a scale that exceeds the needs of the individual family, but what scale is viable? For example, a team of four oxen, using a single plough, can till about 4 hectares of land in a season, and each ox will require about 4–5 hectares of pastoral land for grazing. A farmer can acquire the necessary oxen in two ways. One is to breed them, but then about 15 livestock units are required in order to ensure a ploughing team of four oxen and, here, one is talking of 60–75 pastoral hectares needed for grazing purposes. The other alternative is to buy the four oxen, which presently costs about R6,000 in the Eastern Cape. Unfortunately the output achieved on the farm sizes envisaged here would not make the purchase possible.

Since apparently 90 per cent of the ploughing in the homelands is done by tractor, it is more interesting to explore this option. Tractor services in the homelands have been publicly provided and have a poor record: the tractors are poorly maintained, often inoperational and, when operational, a considerable proportion of their time is spent travelling between plots. This arrangement is now recognized as a failure and privately supplied tractor services are currently being initiated. The importance of viable private sector services becomes evident when one considers a further alternative – small tractors owned by individual farmers. For this to be economical the farmer needs about 100 hectares of arable land, depending on the quality of the land and climatic conditions. One hundred hectares of arable land is viable in that it can sustain a small tractor and produce an income close to R7,000. This alternative is extremely limiting. For instance, if (as is claimed) 12 per cent of the farming area of the four magisterial districts shown in Table 6.2 is arable, only 84,431 hectares of a total 703,592 hectares are arable. This would enable a mere 844 families to settle on arable land.

On the balance of the land one could establish livestock ranches, but a ranch with 100 livestock units would require 500 hectares and would produce an income of only about R6,500. Another 1,240 families could establish small ranches on the pastoral land.

There is no doubt that the figures can be played with. For example, if the rental of tractor services by entrepreneurs proves successful it will be possible to settle households on smaller farms. Even so, the implications for urbanization policy are unlikely to be marked. The issue is not the accuracy of one's calculations – these are at best hypothetical – but whether through the redistribution of land one could settle a sufficient number of people on the land to significantly affect rural–urban migration. The data are not promising. It is chastening to consider that, when farms in Region IIB of Zimbabwe (the high-potential Sandveld area) were subdivided, there were typically markedly fewer small farms than there were labourers.[15]

In contrast, Hattingh's estimate of the optimum farm size in South Africa for maize is about 700 hectares, depending on numerous underlying variables such as the soil, the climate and the farmer's ability. On the other hand, he found no limit to the potential optimum size for a cattle ranch. In his view, the major constraint on the size of a farm is the farmer's ability to manage the farm and to supervise the labour. For maize, this led to the size noted but, for cattle, the constraint was negated by the type of farm and the much lower level of labour input.

Conclusion

Is it desirable to break up large farms into smaller farming units? Weiner *et al.*'s (1985) evidence raised some doubts but, as long as most rural blacks are oriented to migrant labour and lack the skills and motivation to undertake farming, the redistribution of land and the creation of viable small farms in the white corridor do not promise a diminution of urbanization pressures from the adjacent homelands. They are, in addition, likely to lead to a reduction in the agricultural contribution to the economy. It is surely preferable to improve the ability of small farmers to compete with large farmers through access to capital and membership of cooperatives, for example, and then to let market forces influence development. In the meantime, though, it seems reasonable to hypothesize that for the foreseeable future the bulk of the country's future agricultural output will be the product of large farms.

Nevertheless, there is no *a priori* need for a uniform system of land tenure – my view is that one should accept a diversity of forms of tenure. However, if one also believes that the majority of farms outside the homelands will remain privately owned, the problem arises that one has failed to address the plight of farm labour on large farms. Their situation in respect of wage levels and their working and living conditions is, justifiably, something of a scandal. The two means of redressing this situation are the unionization of farm labour and government legislation. Organization of farm labourers is very difficult owing to their dispersed location in small numbers and also to their dependence on the farmer for accommodation and other amenities. Farm labour is therefore very exposed to repressive action on the part of the farmer and, while some company-owned farming units are willing to deal with agricultural unions, the rest are adamantly opposed to them. The government should afford agricultural unions the same rights available to labour unions and should institute minimum wages as well as regulations pertaining to, for example, the access of farm labourers' children to rural schools. However, neither action should be seen as part of an urbanization policy – their effect would be to make rural labour more expensive and, in doing so, they would accelerate the displacement of labour from the farms.

Urbanization and the Transkei

A major unknown in the urban growth of the PWV and other large centres is the response of the inhabitants of the homelands to the progressive removal of obstacles to their urbanization. This part of the chapter consists of an examination of urbanization pressures emanating from the Transkei and, also, some thoughts regarding likely urbanization within the Transkei. I can do no more than speculate regarding the political circumstances that will provide a context for the urbanization process. The Transkei's independent status will no doubt be redefined, but its communal form of land tenure administered by chiefs is unlikely to be changed over the short term. I imagine that an ANC-led government would be rather perplexed about what to do with the homelands and, having reasserted central authority, would move rather slowly. Thus, in so far as the material circumstances of the majority of homeland residents are concerned, I do not foresee much difference arising from political change.

This enquiry starts with the Transkei's economy – can it sustain a larger urban population? The answer, simply, is that there are no prospects for retaining a sizeable, productively employed population in the homeland. There is a complete imbalance between the Transkei's development potential and the growth of its labour force; and industrial decentralization to the Transkei, which is also briefly examined, does not even begin to represent a solution to Transkei's growing unemployment.

Consultants in the Transkei have argued that the development of its towns can help to relieve the homeland's problems and, for this reason, it is desirable to examine the supposed relationship between towns and development. I do this in two stages: first, through a theoretical review of the developmental role attributed to towns and, secondly, through an assessment of the actual role of towns in the Transkei.

In general, the supposed contribution of urban development varies according to one's notion of the development process and whether it encompasses, for example, 'redistribution with growth', 'basic needs' or the more recent World Bank focus on 'accelerated growth'. My review argues for a relative emphasis on the accelerated growth thesis and for extremely modest claims for urban development.

The planners' presumptions are in sharp contrast to what is happening to the towns. (Urban centres within the Transkei are shown in Map 6.1 above.) Trends in population growth, and explanations for them, bring one back to a pointed reality. The rapid growth of peri-urban areas and rural settlements is one means by which the Transkei's population is attempting to cope with its poverty – unfortunately this does not have a marked impact on economic development opportunities in the Transkei.

Peri-urban growth rates in the homelands often exceed 10 per cent per annum. Thus, lastly, I look at public sector policy and the supply of social services. This is a crucial issue, for, when people join peri-urban or rural settlements, they are joining communities where there is usually no capacity for providing services. (Similarly, in contrast to the government's claims that industrial decentralization takes jobs to where the people are, industrial decentralization also provokes agglomerations of people in the centres least able to afford them.) In so far as there is an answer to this issue, it appears to consist of community mobilization and financial aid to obtain the services the communities most want.

The economic crisis

The Transkei's GDP is shown in Table 6.3. It predominantly comprises agricultural output, most of which occurs in the subsistence sector. Agriculture does not offer a significant development path for most of the inhabitants of the region. For what it is worth, the Transkei government's policy is directed at creating a class of 'master farmers', with the result that there is enough land for only about a third of the rural population (Transkei 1983).

The largest sectors in the market economy are the service sector and the public sector. Abedian (1983) has calculated that 40 per cent of the Transkei's GDP originates in South Africa, from, for example, budgetary assistance and customs and excise payments. If one therefore takes 60 per cent of the total GDP of R1,458,250,000, and divides it by the Transkei's population in 1983 of 3,250,000

Table 6.3 The Transkei's GDP, 1985 (%)

Type of activity	% in non-market economy	% in market economy	% of total
Agriculture, forestry and fishing	87.9	12.1	18.2
Mining and quarrying	—	100.0	0.1
Manufacturing	11.5	88.5	7.0
Electricity, gas and water	96.4	3.6	10.0
Construction	27.4	72.6	6.2
Commerce, catering and accommodation	—	100.0	16.5
Transport, storage and communication	—	100.0	2.9
Finance, insurance and real estate	34.2	65.8	5.5
Community, social and personal services[1]	—	100.0	33.7
Total	30.0	70.0	100.0
Total GDP			R 1,458,250,000

[1] In declining proportion: public administration, education services, health services and other.

Source: Development Bank of Southern Africa (1987).

persons less 420,000 migrants (Transkei 1983), the outcome is a per capita GDP of R309. In terms of intrinsic productive capacity, the Transkei clearly is in a most difficult position.

The employment shortcomings in the Transkei are also well known. In 1983 the Transkei's total labour force was 1,100,000, of whom about 725,000 were male. Given that migration is largely a male phenomenon and given, also, that 91 per cent of the Transkei's population is rural and that women play a central role in agricultural production, most employment estimates tend to be rather 'heroic' (the term is Muller's) in respect of their assumptions regarding female labour. The figures regarding the smaller presence of women in the labour force should therefore be treated with some caution.

The location of the labour force in 1982 is shown in Table 6.4, which reveals that only about 19 per cent of the formal sector labour force is employed inside the Transkei. Unemployment was then just short of 20 per cent; now it is certainly more.

Tables 6.4 and 6.5 together again demonstrate the Transkei's extreme dependence on migrant and commuter opportunities. Migrant labour occupied two-fifths of the labour force and, in 1985, its earnings were R1.66 billion. Remittances of 40 per cent

Table 6.4 Structure of the labour force, 1982

	%
External	41.3
Formal	19.2
Unemployed	18.4
Subsistence	14.7
Informal	6.4

Source: Muller (1984: Table 3).

Table 6.5 Sources of rural household cash income, 1982

Annual income (R)	H'holds %	% contribution to household cash income			
		Wages	Pens.	Remit.	Prod.
<500	23.9	10.65	19.40	76.05	2.90
501– 1,000	26.4	12.10	14.30	71.10	2.50
1,001– 1,500	13.2	15.36	17.11	65.75	1.78
1,501– 2,000	6.7	26.52	21.40	48.08	4.00
2,001– 3,000	6.9	37.67	13.89	46.17	2.27
3,001– 4,000	6.5	74.40	4.24	19.86	1.50
4,001– 5,000	5.4	80.61	4.22	13.22	1.95
5,001–10,000	7.3	69.90	4.43	22.95	2.72
>10,000	3.7	83.05	8.16	6.08	2.71

Source: *Transkei Profile*, (1985: no. 1); cited in Muller (1987b: 60).

of these earnings enable the workers' families in the Transkei's rural areas to survive (Abedian 1983). Table 6.5 shows that remittances constitute over 60 per cent of the cash incomes of more than 60 per cent of the rural households and, as the Transkei's population is predominantly rural, this is a very significant figure. Only 23 per cent of the households earn the bulk of their livelihood from wages. Agricultural production is understated in the table as it ignores income in kind, but the table does usefully reveal the minimal extent to which all households are engaged in production for the market – the highest contribution of agricultural cash income to any level in the range is 4 per cent.

INDUSTRIAL PROSPECTS

Could industrial development make a difference? The question has two parts, as described by the efforts of the Transkei Development Corporation (TDC), responsible for administering South Africa's industrial decentralization programme in the Transkei, and the Transkei Small Industries Development Organisation (Transido). Different sets of logic are appropriate to an assessment of the decentralization programme. In terms of a South African view, there is general agreement that the decentralization programme has either to be abandoned or substantially revised. A Transkeian's perspective is different because the Transkei represents an inefficient industrial location that would, in large part, not otherwise be used were the subsidies not present (Addleson *et al*. 1985) – contrary to Muller's (1987a: 20) unique view that the 'Transkei is centrally located' and that industrialization in the Transkei is viable. A cost-effective industrial location policy for South Africa would offer little gain for the Transkei, which thus has reason to want to persist with the decentralization programme.

In respect of the programme, the real value of Transkei's manufacturing output has declined since 1982 (Muller 1987a), which is particularly notable since the new concessions, which prompted a rapid increase in decentralization elsewhere, were introduced then. Transkeian locations are at a competitive disadvantage because the Ciskei's industrial development points are closer to East London and are on the main railway line to the Witwatersrand.

By 1986 only 16,244 wage labour opportunities were credited to the decentralization programme. The relative contribution of this employment to household living levels in the Transkei is evident when one discovers that the average monthly wage paid by the manufacturing sector in 1985/6 was R120 and that the total annual wage bill, over this period, was equal to about 1 per cent of consumer expenditure in the Transkei (Muller 1987a).

Industrial decentralization plays a minimal role in the Transkei's economy and has little potential to play a greater role. It is far more interesting to speculate on how a secondary city strategy would affect the Transkei. The closest secondary city to the Transkei is East London. It is hard to dispute the view that the Transkei's most economically feasible option would consist of industrial expansion in this city coupled with free migration, alongside attempts, within the Transkei, to transform agriculture.

In regard to small-scale industry, Transido's efforts appear promising because of the apparently low cost of the jobs 'created' and the fact that the enterprises aided are not ephemeral concerns but represent, rather, persons who will stay in the Transkei and who will exploit local linkage opportunities when this is possible. Enthusiasm in respect of small-scale enterprise should not, however, disguise the fact that it is of distinctly limited potential – most of it is, inevitably, in commerce and personal services. In the case of small-scale industry, a recent survey revealed the following incidence: knitting and sewing (64.6%); metal working (13.1%); carpentry, furniture-making, upholstery and fibre-glass working (10.9%); textiles and leather (6.9%); construction (2.0%); and other (4.0%) (Hofmeyr 1985). Close to 99 per cent of the industries had fewer than ten employees and 95 per cent had fewer than five. The 'average operating surplus per enterprise' was about R161 per month, which is better than wages in decentralized industry but, of course, applies only to the entrepreneur – wage levels for the others are likely to be much lower.

The primary constraint to Transido's operations is not capital so much as the supply of potential entrepreneurs. The major problem restraining existing entrepreneurs deals with marketing and the size of the market.[16] This finding is paralleled in a World Bank report (Page and Steel 1984) in which the authors state that the constraints to the growth of such enterprise consist not of supply-side problems, such as shortages of capital (which are predominantly supplied through savings and loans from families and acquaintances), a lack of premises or the absence of business skills, but rather the extent of the market.

In effect, growth is contingent on improved incomes for the mass of the low-income population. The growth in consumer expenditure between 1979 and 1984 in the Transkei averaged 22.1 per cent per annum. In itself, this is very promising but it must be remembered that this high rate of growth resulted from its small base, the burgeoning public sector and the growth of migrant incomes in South Africa. In other words, it is contingent on migrant and public sector incomes, which derive, either directly or indirectly, from sources in the rest of South Africa whose continued growth potential is limited.

It would, in addition, be a mistake to relate the growth in consumer spending to a proportionate increase in demand for the output of small-scale enterprise. As the Transkeian market grows, so it becomes increasingly attractive to large South African enterprise. It is precisely the small size of the Transkeian market and its inaccessibility that ensures a niche for the Transkeian entrepreneurs. This is especially evident in the case of Flagstaff, whose unusually high level of manufacturing activities is explained by Hofmeyr (1985: 4) as 'due to its isolated situation, well away from other major sources of supply of common goods and connected to them only by relatively poor roads'.

The supposed developmental role of small centres[17]

All perceptions of the development process hold that small centres can support rural development by providing markets, being the site of processing and other informal industries, providing agricultural supplies and repairs, supplying social and commercial services, being part of a broader network of roads, electricity and water . . . and so on. But the perceptions vary in their relative emphasis on the economic or social role attributed to the towns: that is, whether centres engender economic development or should be seen primarily as efficient sites for the location of social services. There is a 'chicken or egg' aspect to this position, as one could alternatively say that it is the transformation of the rural economy that will make an expanded economic role possible for medium and small towns. For example, one cannot but be impressed by the fact that the Transkei Agricultural Corporation frequently bypasses the towns for sites when supplying agricultural inputs and when coordinating the marketing of produce. It is less than clear that towns contribute to rural development, as opposed to improved agricultural incomes making it possible for the towns to grow.

These contrasting views should be obvious from economic models with titles such as 'basic needs' and 'accelerated development'. There is consequently little agreement regarding the role of public policy and the nature of public investment programmes. While the debate may be fairly heated, as in the case of the Transkei (Osmond Lange *et al.* 1982; Tapscott *et al.* 1984; May 1985), it remains true that the theoretical basis of the debate is fairly vacuous. It is as well, at the outset, to clarify the limited theoretical grounds for policy intervention.

Initially, in the 1960s, the perception of development economists was that the development process in the rural areas consisted of transforming subsistence production into production for the market. In this view, dispersed urban centres provided markets, and intraregional transport and marketing systems provided access to markets in other larger centres. Urban growth was also taken to facilitate the diffusion of new ideas, which, in turn, supposedly helped extension agencies and the market to introduce innovations in production and, through exposing the peasantry to new goods and services, caused them to produce for the market in order to expand their consumption possibilities. At the same time, awareness of urban opportunities and urban–rural differences in living levels would promote urban migration. This was considered desirable because it would enable those remaining behind to accumulate larger land holdings and also because economists were, at this stage, concerned about ensuring an adequate supply of urban labour. (It was only later that the imbalance between the growth of the labour force and creation of employment was realized.) An urban hierarchy, based on differentiation among urban centres owing to dissimilarities in the range and threshold of their products, was supposed to arise.

In the case of the Transkei, the planners' delineation of urban hierarchies reflects the different perceptions regarding the social or economic role of the towns. Consultants in the Transkei have held that the centres constitute 'focal points needed to stimulate socio-economic activity and innovation' (Osmond Lange *et al*. 1983b: 2), but it has also been argued that the towns lack any economic potential and that urban hierarchies are primarily necessary in order to coordinate the location of public investment.

The reason for the latter view is that most of the towns are declining. Agriculture in the region is very poorly developed: commercial agriculture is marginal, and subsistence agriculture, while a prominent item in the Transkei's GDP, involves only minimal interaction with the market via the towns. The poverty of the rural inhabitants – an average rural household income of R1,152 in 1979 – also does little to contribute to a commercial role for the towns.

The public sector, not agriculture or manufacturing, has been predominant in determining the role of the towns. This is evident from the fact that public sector output (across all sectors) contributes more to the Transkei's GDP than subsistence and market agriculture combined, and almost four times as much as manufacturing (Abedian 1983). It may be argued that public expenditure does not denote urban expenditure but, in 1983/4, 73 per cent of the public sector budget was spent in urban areas (Thomas 1984). Unhappily, there is every reason to look to the location of public facilities and employment, rather than to economic advance, for an understanding of the role of towns in the Transkei.

This tendency to accept the predominant role of the public sector is especially clear from the Spatial Development Plan for north-east Transkei (Osmond Lange *et al*. 1983a). The consultants actually distance the plan from production issues relative to those that have to do with consumption. Thus, '"non-issues" include problems like widespread poverty, unemployment, migrant labour, access to arable land, low productivity of farmers on betterment schemes . . .'; and 'priority fields of action . . . should relate to the provision of public sector goods such as infrastructure, education, health services, administration and means of communication' (p. 3). In other words, little opportunity is seen for greater linkage between towns and economic advance. This represents a position of considerable despair for regional planners, but does not do the planner out of a job as there are still calls for an urban hierarchy to direct public investment.

The existing and prospective hierarchy specified in the Transkei's 1983 White Paper is shown in Table 6.6. The hierarchy depicted for 2003 is premised both on relatively 'hard borders' and on successful internal planned development. Neither assumption is correct, with the consequence that the urban population estimated for 2003 is three times that anticipated by Muller (1987b). (The overestimate is, however, only about one-and-a-half times the likely population if peri-urban growth is taken into account.)

Urban hierarchies tend to represent wishful thinking by planners. In the case of the Transkei White Paper, for instance, there is an unclear relationship between this hierarchy and the actual number of centres identified by Osmond Lange *et al.* (1983b: 167–8) who present the following hierarchy:

A *Capital City* – Umtata
B *Major Towns/Service Centres* – e.g. Idutywa, Mqanduli, Engcobo (about 10–12)
C *Minor Towns/Service Centres* and intermediate places with a high concentration of population or economic activity – e.g. Kentani, Willowvale, Elliotdale (about 20–30)
D *Consolidated Villages* (at least one per Administrative Area where 'betterment schemes' have been implemented – about 500–600); [and] *Traditional Villages* (at least one per Administrative Area where 'betterment' planning has *not* been undertaken – about 400–500).

It was originally thought that the formation of an urban hierarchy was fairly automatic – the outcome of market forces and migration processes. The importance attached to the hierarchy is evident from the view of an especially prominent scholar of the time:

Table 6.6 Existing and prospective urban hierarchy (White Paper, 1983)

| | No. | Population per centre | | Total 'urban' Pop. 2003 |
		1983	2003	
Umtata	1	70,000	200,000	200,000
Butterworth	1	30,000	50,000	50,000
Other regional centres	4	3,000	50,000	200,000
Market towns	20	1,000	15,000	300,000
Rural nodes	90	400	5,000	450,000
Total				1,200,000

Source: Transkei (1983: 20).

[its] developmental role involves the simultaneous *filtering down* of innovations that bring growth down the . . . hierarchy and the *spreading* of the benefits accruing from the resultant growth, both nationally from core to hinterland region, and within these regions from their metropolitan centres outwards to the intermetropolitan periphery. Regional inequities arise in this scheme because the income effect of a given innovation is a declining function of time and is also subject to a threshold limitation – a minimum size of region – beyond which diffusion will not proceed. As a consequence, the lowest levels of welfare are found in areas peripheral to small centres in outlying hinterland regions.

(Berry 1972: 340–1; emphasis in original)

In this scheme of things, policy interventions consisted of filling in 'gaps' in the hierarchy that 'blocked innovation diffusion' and, to a limited extent, also of promoting the development of the small centres in outlying regions. Notions of urban hierarchical innovation diffusion may be helpful when they provide a coherent means of ordering and locating investment in public infrastructure and services. However, the theory is taken too seriously when intrinsic properties of innovation diffusion – of 'bringing growth down the urban hierarchy' – and of urban development are thought to provide a means of promoting rural development. Circumspection is therefore appropriate when one reads that a rural service network 'based on a hierarchy of functions . . . will provide the channels for encouraging development in rural areas' (Osmond Lange *et al.* 1983b: 2).

The slowness of development in creating 'trickle down' effects among people and regions and in creating urban hierarchies has not led to disillusionment with the urban investment programme. On the contrary, as one moves from the 'redistribution with growth' to the 'basic needs' initiatives, increasing attention is paid to the need for such programmes. Subsequent uncertainty regarding the automatic creation of lower-order centres led to calls for the creation of villages, holding centres and, especially popular, 'rural service centres'. Thus in the Transkei (1983: 8) a government document states:

If the people working on the land are to earn their living from the land, they will need markets in which to sell their output. This will be one of the primary functions of smaller towns. If the agricultural population is to produce enough food to feed the required three to four times its number it will need inputs from larger centres than those where the farmers now live. To achieve this aim Transkei requires an effective policy instrument and it is proposed that a Service Centre approach be implemented . . . this should result in the establishment of a hierarchical network of physical bases from which to deliver public services; agricultural support services to increase productivity with a view to transforming subsistence into market orientated farming; the provision of basic infrastructure, i.e. feed roads and water for domestic use; stimulate rural land, industries. These inputs require commitments in terms of capital and investment – the location and spatial arrangement of which is important. Rationally they should be located in centres that are accessible to the local population. In this way a system of rural towns, or service centres, can be established on a hierarchy of functions. The resulting network will provide the channels for encouraging development in the rural area.

Here service centres are taken to have an active role in promoting agricultural development and, unlike the earlier pessimism, they are not seen simply as the sites for public expenditure and services. The question one has to ask is whether this is a realistic position.

The fashionability of rural service centre approaches also follows from the popular 'integrated rural development' programmes of the 1980s. These consist of soil conservation, general extension services, roads, nutritional education, schools, credit, and so on. The proliferation of public sector inputs led to a search for optimal sites, a need that rural service centres served well. As it turns out, integrated rural development approaches are usually beyond the capacity of a Third World country to duplicate in more than one or two politically favoured regions (Lele 1975), and they occur in seeming disregard of resource constraints and the need to earn a return on one's expenditure (Eicher 1986). It is therefore not surprising that alternatives are now being sought for these programmes.

In international consulting circles, and in institutions such as the World Bank, there is a return to economic fundamentals. While one might object that this move was prompted by the ideological leanings of the Reagan White House, it is also a response to the development and debt crises confronting African countries. Thus the Bank (IBRD 1984: 1) 'stresses the better use of investment – . . . Making the most of investment requires not only appropriate pricing policies, but also adequate management capacity in the government, . . . it requires a more active role for non-governmental institutions and for the private sector'. Needless to say, the change in orientation has been subject to considerable debate (Green and Allison 1986; Shaw 1986; Please and Amoako 1986).

Accelerated development with regard to agriculture constitutes a call for higher producer prices for farmers, exchange rate adjustments that reinstate positive returns to exporters, a loosening of controls on input prices, efficient marketing and ensuring the availability of inputs and consumer goods (incentives for production). The associated implications for settlement policy are that:

- the focus is on investment in areas where the return is greatest – for South Africa as a whole these are usually large urban centres;
- in rural areas the emphasis is on agricultural modernization, which implies the dispossession of subsistence farmers; and
- investment in areas such as the Transkei is justified only by profitable opportunities and, in their absence, the region should be allowed to decline. The people can move to where jobs can be found.

Unlike the modernization theory of the 1960s, which allowed a 'role for policy in promoting an integrated hierarchical system of settlements to assist innovation-diffusion, the accelerated growth approach *implicitly* holds [that] the settlement system will move towards the "best" form automatically through market forces: artificial attempts to adjust the hierarchy before the market is ready simply represent wasteful investment' (Dewar *et al*. 1986: 113).

For town and regional planners, this is a particularly disconcerting view. Two decades of international experience of attempting to promote growth centres and rural service centres have led to the conclusion, simply, that they do not work. The position now is that the issue is more one of managing the growth of the large cities than of attempting to disperse or restrain that growth. If, for political reasons and not those of efficiency, governments wish to redirect some employment creation, then 'secondary cities' are now the preferred location. The Transkei, however, has no secondary cities. In a situation such as that of the Transkei, the World Bank view and decades of experience leave one with an impenetrable dilemma.

What is happening in the towns?

With regard to urban growth in the Transkei, Table 6.7 shows a rapid rate of growth for the three largest centres. It is quite easy to guess at the reasons for their growth: the expansion of the civil service in Umtata, the industrial development point status of the three centres and relocations into Ezibeleni.

The other towns consist of 'medium towns', such as Engcobo and Idutywa, and 'small towns', such as Elliotdale and Willowvale.

Table 6.7   The distribution of urban population, 1960–1985

| District | % urban population | | | | Growth rate 1960–85 |
	1960	1970	1980	1985	% p.a.
Umtata	36.7	41.2	32.7	38.9	6.4
Butterworth	7.1	4.6	21.2	18.0	10.2
Cacadu/Ezibeleni/					
Illinge	2.8	2.1	14.8	21.0	14.9
Medium towns[1]	27.4	31.1	18.2	10.7	2.2
Small towns[1]	26.0	21.0	13.1	11.4	2.7
Total	100.0	100.0	100.0	100.0	6.2

[1] Medium towns were defined as having a minimum urban population of 2,000 in 1981. They included: Engcobo, Idutywa, Kwabhaca, Tsolo, Umzimkulu, Umzimvumbu, and Xalanga. Small towns comprised the balance of the urban population and fell into 18 districts.
Source: Muller (1987b: 47).

Table 6.7 points to a gloomy picture regarding their development. Between 1960 and 1985, the population of medium towns declined from being 27.4 per cent to 10.7 per cent of the Transkei's urban population. Their growth rate over the same period was 2.2 per cent per annum, which is below the rate of natural increase.

The low growth rate of the towns is unexpected since a number of factors would seem to precipitate growth. Most importantly, only about 20 per cent of the rural population wish to farm. If this is coupled with the increasing number of the landless, the higher cost of food and transportation in rural areas and the failure to utilize a large proportion of the Transkei's arable land, then it seems reasonable to expect rapid urban growth.

One explanation of rural immobility is suggested by Baskin (1984: 12–13). After first noting that land is 'of only tertiary importance as a household income earner', he continues that it 'is within the notion of material benefits and the future of the children, that the answer to the question can be partially found. There is a common belief, correct or incorrect, that in future years land will be impossible to get. . . . [Further], as scarcity increases so will the value of the land increase . . . [and] the belief that in future years, agricultural productivity will increase as a result of new technologies'. To leave the land is to give up one's rights to the land and it would be irresponsible to do so.

Furthermore, in the rural areas, relationships within the extended family and the clan provide support during periods of hardship. Material survival may be further contributed to by the rural residential site having a small plot for growing crops, although, increasingly, 'no new residential sites are available any more. In practice, arable allotments and residential sites are now passed from a man via his widow, to his (usually) eldest son' (De Wet 1987: 461).[18] In addition, so long as migrant remittances are forthcoming, there is not the pressure to urbanize in order to seek out alternative livelihoods.

The other side of the coin deals with opportunities in the towns. Surveys reveal that the primary reason for a rural household's decision to relocate to the urban areas involves the search for jobs, but the employment opportunities are unlikely to continue to increase as in the past. The major source of growth has been the expansion of Umtata's civil service, but transfers from South Africa are being cut back and the jobs that 'needed' to be filled are in large part occupied. It seems very unlikely that, under the hypothesized unitary, democratic government, there would be much further attention to building up the administration of the homelands. As noted, the growth of employment in Umtata, Butterworth and Ezibeleni has also been contributed to by their industrial development point status, but this programme is likely to be short-lived in its present form. In addition, it is improbable that a democratic government would base the selection of sites for any possible future decentralization programme on locations that served the ends of apartheid. In most other respects, the towns lack an economic base. This is especially the case for the medium and small towns. Some, like Idutywa, retain a commercial vitality but, on balance, the towns are declining.

The lack and the cost of serviced land in the towns is yet another impediment to urbanization: use rights for a residential plot in the towns cost R1,000, the transfer of ownership costs another R1,000, there are service charges, and urban building regulations increase the cost of one's residence. In contrast, sites in rural villages cost a sheep as a gift to the chief and, perhaps, another smaller present to a headman in order to gain access to the chief. Besides, the gift occurs once only. Peri-urban sites are sometimes more expensive, involving, perhaps, a R10 per annum fee, but the present might be less onerous, something like a bottle of brandy.[19] It seems that households actually prefer peri-urban locations and that, for most people, towns represent a second, even a third, choice.

In the case of peri-urban growth, Zones A and B around Umtata, indicated in Table 6.8, grew at 17.4 per cent and 15.3 per cent between 1980 and 1985 (Rosmarin, Kriek, *et al.* 1986; cited in Muller 1987b: 50). This very fast growth rate is markedly higher than the 1970–1980 growth rates of 5.6 per cent and 4.9 per cent, respectively. The difference in the growth rates between the two periods causes one to be ambivalent regarding the data.

Table 6.8   Estimated peri-urban population by urban type, 1985

Urban type	Urban population	Peri-urban population
Umtata	57,796	44,283 (Zone A)
		43,634 (Zone B)
Butterworth	26,776	49,434
Cacadu/Ezibelini/		
Illinge	31,112	0
Medium towns	15,853	39,633
Small towns	16,976	25,464
Total	148,513	202,448

Source: Muller (1987b: 51).

Yet, even if an overstatement is involved, the data none the less point to increasing numbers of people lacking rural resources or migrant remittances that would enable them to stay in the rural areas. Given these numbers, it would be irresponsible not to plan for a future that is revealed in the enormous peri-urban growth and the occasional reference to the rapid growth of a number of villages. In the latter regard there have been reports of villages up to 6,000 persons, larger than the medium towns. Table 6.8 shows that the Transkei's peri-urban population already exceeds its urban population. In the case of medium towns, it is double the urban population. It would therefore be incorrect to view an urban development policy as referring solely to officially designated towns.

Peri-urban growth, however, is not welcomed by the Transkeian government. Thomas (1983: 238) writes:

policy makers at government and local authority level are extremely reluctant to allow this process of rural–urban shift, fearing in particular the danger of (politically destabilising and aesthetically unappealing) unemployment and squatting in or around the towns. Tight controls on squatting and hawking, limitations on the land made available for (low cost) housing and strict (often antiquated) town planning regulations are steps to prevent rapid urbanisation. In addition, government emphasises tribal structures and the duty of (unemployed) people to go back to 'their village' as a first resort. Implicitly it is assumed that tribal and family ties will function as 'unemployment insurance'. . .

Elsewhere Thomas (1984: 4) reports that 'Small district towns . . . are virtually starved of funds, so that they limit their activities to a bare minimum of urban infrastructure services'. Dewar *et al.* (1984 and N. Nattrass (1984) describe a similar reluctance to facilitate the growth of the towns.

However, it may be that, as far as the Transkeian government is concerned, the above comments are out of date. Interviews with government officials pointed to official encouragement of the population and economic growth of the small towns. It is another thing, of course, whether this encouragement extends to the supply of capital and technical aid – only time till reveal the strength of the government's policy switch. The interviewees disagreed regarding the perceptions of the town administrations on this issue, with one well-placed informant still seeing some reluctance to admit urbanization while others put the problem down to technical and administrative shortcomings and the lack of capital – all exacerbated by out of date legislation. The government is held to be promising changed standards and financial and technical aid.

Nevertheless, there is little reason to expect that a future government can make land and housing affordable to the Transkei's potential urban population. The demand is too great and the government's resources are too few. In the latter regard, for example, the revenue base of Butterworth is about R1.5 million and that of the next 20 towns is about R1.6 million.[20] Experience, moreover, does not lead one to believe that governments are, in fact, concerned with making serviced sites or housing available to lower-income groups. Around the world, governments have subsidized the housing of middle- and upper-income groups, especially when they are government officials. The Transkei government has not been, and is unlikely to be, an exception.

Future uncertain migration patterns

The labour force is growing by about 30,000 persons a year and the consequence is an anticipated labour force of 1,750,000 persons by the turn of the century, of whom about 1,125,000 are expected to be male (Transkei 1983). The Transkei cannot foreseeably meet the need for employment, which, ideally, would be based on the growth of the industrial and agricultural sectors and be accompanied by the emigration of many families.

As far as agriculture is concerned, the Transkeian government has stated that 'A rapid deployment of resources directed at the creation of a class of full-time farmers, who would produce crops and livestock products for sale on local and export markets, is needed if agriculture is to generate sufficient new employment opportunities' (Transkei 1983: 9). As noted, the government reckons that the employment created will suffice for a third of the labour force and its goal is to sustain the farmers at their *present* income levels. In broad terms, therefore, agricultural policy is not, indeed cannot be, designed to hold people on the land. Emigration is the obvious alternative.

The future movement of the Transkeians is obviously very important, but in this respect one is speculating with uncertain trends. For example, Giliomee and Schlemmer (1985) state that surveys reveal that about 25–30 per cent of migrants would like to move to the cities and give up their ties to the rural areas; Moller (1988) explains this by the fact that migrants' intentions are primarily determined by where they find greatest security. She suggests that the removal of barriers to migration will not unleash a large-scale move to the cities, but also that this is likely to be a generational matter and that, in the future, urban migration will be more pronounced. I agree – as I see it there is an increasing lack of security in rural areas. Landlessness figures of 30 per cent and 40 per cent have already been mentioned, and the situation is rapidly worsening. For many there is not a rural option. It has been possible for people to reside in rural areas, but without sufficient land or cattle to make a living, only because of migrant remittances – but migrant options are rapidly decreasing. On the one hand, whereas the South African economy needed to create about 2.5 million new jobs between 1980 and 1987, it in fact lost 150,000 jobs (*Business Day*, 5 July 1988). On the other hand, between 1982 and 1987 the share of migrant labour in the mines declined from 70 per cent to 40 per cent. In addition, new mines, which in the past might have created about 30,000 jobs, are now more capital intensive and add about 6,000 positions to the labour force. People are being forced from the land and this is especially so for the younger generation. Future migration to South Africa's cities is inevitable; in the meantime, peri-urban growth within the homelands is occurring very rapidly.

Social services in peri-urban areas and rural settlements

It would be difficult to defend the spatial allocation of social expenditure in terms of a predetermined urban hierarchy. The growth patterns of urban areas, peri-urban areas and rural settlements are not following the planners' hierarchical prescriptions. Would it not be a great deal simpler to conduct periodic aerial surveys to map the emerging areas of concentrated settlement, and to locate public expenditure on schools, water and clinics accordingly?

This suggestion does not prohibit an attempt to influence the form of development – it costs a great deal less to lay a water pipe down a straight line rather than to have it follow the bends characteristic of paths in informal settlements. However, expensive consultants hired to advise government departments what types of investment to undertake in a specific level of centre may be worse than having no advice at all. First, people move to centres where their best opportunities lie, and predetermined investment locations are unlikely to be of much help in guiding migrants. Unless growth is occurring in areas where it is particularly inappropriate, rather let the government react to locations that emerge 'naturally' and try to accommodate growth in as economically efficient a manner as possible. Secondly, the actual investment prescribed within a hierarchy – 1,000 people equals a primary school – may not reflect the priorities of the people, which may be the protection of a spring or an improved access road. The participation of the people in determining both the nature and the location of the investment is critical.

Community participation is intrinsically desirable, but there are additional reasons for it in the Transkei since the settlements and peri-urban areas usually fall under a chief or headman who lacks the administrative, financial and technical capacity necessary to service the settlements. Governments of middle-income countries cannot afford to supply all the necessary social services and infrastructure and, in peripheral areas, will likely not see much reason to. Democratic community organizations represent an effective alternative. (It is important that the organizations are democratic because, as soon as community organizations are led by functionaries without legitimacy, participation in community schemes dries up.) This proposal consists of identifying areas of population concentration, encouraging the evolution of democratic community organizations that will generate limited resources on their own, and providing them with some capital and technical expertise. The investments undertaken and the services supplied will, inevitably, reflect budgetary constraints, but will also reflect the priorities indicated by the communities. What I am suggesting is that governments engage in less planning. Were the above suggestions followed, the outcome would be to reach more people at a lower cost.

None of this, however, has much to do with economic development. The issue concerns serving the needs of concentrations of people who will largely live outside the proclaimed towns; the earlier pessimism, noted in respect of a regional development plan that did not pretend that the towns had a significant economic role, has been borne out. The goal is simple and rather desperate: to look after the poor and the unemployed and to help them to take care of themselves.

Conclusion

The conclusion to the material on the redistribution of white farms is rather disappointing. It would have been marvellous had I been able to argue that it was possible to settle people more densely on the land and that there were no costs in respect of economic efficiency. Many might still quibble with my findings; obviously a lot depends on one's assumptions and goals. For instance, if one's starting point is to give back stolen land or to create a more racially equal distribution of rural land holding (with the presumption that this will mostly require the subdivision of farms), then there is inevitably disagreement with my position.

In respect of an urbanization policy, however, the point that most distinguishes farming in South Africa from Third World agriculture and that has the most bearing on the arguments surrounding the redistribution of land is that South African agriculture exhibits returns to scale. If one's goal is to maximize agriculture incomes, then the appropriate scale of farming has exceeded the small farm that Hunt (1984), Berry and Cline (1979) and many others have in mind. It is also striking that the scale of farm necessary in order to provide a household with an income of R6,000–7,000, which has been put forward as the income level necessary to restrain migration, does not promise to settle people more densely on the land. Without attempting to disparage agriculture, South Africa's future is an urban future, and surveys of the people suggest that this is their orientation.

With regard to the Transkeian study, at the outset there was reference to the view that urban policy has more to do with coping with the Transkei's development crisis than with promoting economic development. As it turns out, this seems, in large part, an accurate perception of needs and potential in the Transkei. Some commercialization of agriculture is likely, but this probably will not significantly involve, or be contributed to by, urban centres. Instead, the centres will have to cope with the displaced rural population. These centres, however, represent high-cost locations and, with the exception of Umtata, lack the technical capacity and capital to significantly expand the supply of bulk services.

The implication is that future urban growth will largely be peri-urban and will also occur in the villages. In this light, the predominant issue is organization for the provision of services. The limited administrative capacity, as well as the resource constraints, facing any future government point to the need to promote community organization. This is necessary in order to prioritize the people's needs, to provide labour, to contribute to the cost of services and infrastructure and to administer and maintain those services and infrastructure.

There seems little the public sector can do in order to promote economic development. Some agricultural development is conceivable, but it will displace a significant proportion of the population. With regard to industry, even were the industrial decentralization programme to be retained, there is unlikely to be much industrial investment. A local blossoming of small-scale enterprise is also unlikely if, as I expect, migrant labour increasingly settles outside the Transkei – those who most contribute to the size of the local market and to the pool of savings and skills will leave.

In so far as planning for the Transkei's future settlement pattern is concerned, considering the small size and minimal economic base of many of the designated towns and the formation of new towns, such as Mtojeni, as well as peri-urban growth and the growth of many villages, it would probably be correct to argue that what is occurring is the creation of a new urban hierarchy in the Transkei. My planning suggestions are far simpler than those normally associated with urban settlement planning. The government should set out to monitor where urban growth is occurring, to restrict it only if absolutely necessary and, then, to provide the necessary services. It need not be entirely fatalistic – since Umtata will grow rapidly, the government can attempt to lead growth through its location of infrastructure and services. Again, however, international experience is that people have a rather clear idea regarding where their optimal sites are. Investment – for example, the provision of serviced land as 'catchment areas' – may well be to no avail if it conflicts too markedly with the wishes of the people.

Notes

1 These figures were supplied by D. R. Tapson and N. Vink of the Development Bank of Southern Africa. It would be misleading, though, to attach too much significance to these measures as only about 12 per cent of the homeland population's food requirements is produced there. Homeland economies are primarily dependent on migrant remittances, commuter incomes, and employment generated in the public sector through fiscal transfers from the South African government.

2 Professors T. J. Bembridge and D. R. Tapson (who later joined the Development Bank).

3 Dr Chris De Wet, Rhodes University, Grahamstown.

4 In conversation. Mr H. Hattingh is Director General of the Department of Agriculture.

5 In conversation.
6 The rough estimate was provided by J. Ellis-Jones of the Transkei Agricultural Corporation. He added that the degree of utilization of land improved for better-quality land, and vice versa.
7 The point was made during conversation.
8 It is very difficult to determine when land is under-utilized. Land that is conservatively managed may appear under-utilized. Likewise, land left fallow, which it should be every three years, may appear under-utilized.
9 Muller (1987b: 62) reports that it is commonly the case that more than 50 per cent of the households in a village lack cattle and more than about 80 per cent of the households lack sufficient cattle to form a team for ploughing. His presumption is that it takes eight oxen to form a team, but four oxen can pull a single plough.
10 The source is again J. Ellis-Jones. Professors Tapson and Bembridge mentioned figures of 10–20 per cent for the Ciskei, which corroborates Ellis-Jones' estimate.
11 University of York, Centre for Southern African Studies, conference on 'The Southern African Economy after Apartheid', 29 September – 2 October 1986.
12 The data were provided by C. Butcher (1985), who was citing the Zimbabwe government's Central Statistical Office, *1985 Agricultural Statistics*.
13 Mr J. Hayward, former director, Department of Agricultural Technical Services and Extension (AGRITEX), Zimbabwe, 1981–85.
14 In conversation.
15 For example, a tobacco farm will normally be about 1,000 ha in size, of which about 25 per cent will be arable. Land used for tobacco will be rotated so that following one year planted with tobacco it will for three years be planted with grass. Typically about 50 ha will be planted with tobacco and each hectare will require one labourer, which means one labourer plus his family – in all about 300 persons. Another ten labourers plus their families are likely to be employed elsewhere on the farm. Thus the farm will support about 60 male labourers and their families, some of whom might also be employed during peak seasons. When such farms were subdivided each farm was typically about 33 ha, of which about 5 ha were arable. Since each family uses its own labour and seldom employs others, the outcome of subdivision on high-potential land is a considerable reduction in the number of people who gain a living from the land.
16 The source if Mr J. Tregay of Transido.
17 In presenting this material I owe a considerable debt to Dewar *et al.* (1986).
18 The quote refers to conditions in the Ciskei, but they are rapidly coming to approximate those in the Transkei.
19 T. Wanklin of Infraplan, Umtata, provided these figures.
20 ibid.

7 Conclusion

South Africa's current urbanization strategy disregards the needs of the people: it serves the ends of a repressive regime and suffers, especially, from the fact that it is not a product of the democratic process. Therefore, it is all the more unfortunate that, even with the advent of democracy, South Africans will continue to be burdened by the apartheid legacy. For instance, the displaced urbanization and investment in isolated townships, the inadequate housing and services, and the overcrowded homelands are some of the many problems that will continue to weigh heavily upon the country.

The future is more promising, since there are options available whose effects are to relieve poverty. Policies that are intended to overcome the inequitable and inefficient design of the cities; to accentuate the supply of land and housing; to increase the number of jobs in poverty-sticken regions; and to improve working conditions on the farms – all of these constitute ways in which an urbanization policy can reach out to the poor.

It would be a mistake, though, to believe that urbanization policy can, on its own, solve many of the problems deemed to be urbanization problems – many problems are found *in* cities or regions rather than actually being a problem *of* that city or region. Thus, a city government might attempt to relieve the shortage of housing located within the city, but the central problem may actually lie with national economic policy and the slow growth of employment, which together prevent the poor from being in a position to afford their own housing.

One has to appreciate that the role of urbanization policy, and its ability to resolve social crises, is, in fact, modest – one cannot, for instance, realistically expect to house all the people. Both the government and the poor are unable to afford an adequate supply of formal houses and, for both parties, formal housing may not be a priority. As has been seen, the poor often have other, more weighty, objectives – employment, food, schools – and are prepared to settle for a shack. A formal site-and-service scheme or a house grow in importance as the income of the household increases and, if the poor do acquire a free or subsidized site, they are likely immediately to relinquish it through a process of downward raiding.

In the context of pervasive unemployment and, initially, slow growth, the government is also unlikely to direct the economy's resources to housing – especially as economists are uncertain that the returns to such investment would contribute as much to the economy as would other potential investments. Moreover, one should probably be grateful if the government does not make housing its first priority since, when governments do, their expenditure is likely to occur in the form of subsidies to favoured constituencies. Limited cost recovery reduces the potential for future investment, growth and the reduction of unemployment. The conclusion one arrives at is that governments can, and should, strive to ensure housing for the poor and the unemployed, but that they should do this through carefully selected programmes and legislation. If, as the World Bank holds, the bottom 20 per cent might not be housed, this is not a reflection on the government but is the result, rather, of unavoidable economic constraints, which will be alleviated more rapidly though economic growth than through inefficient housing programmes.

Another similar, disciplined, conclusion is appropriate with regard to industrial decentralization. Richardson (1978: 49) writes that while 'measures to raise the incomes of the rural poor are easily justifiable on equity grounds, "tinkering" with the settlement pattern is not an efficient way of doing this . . . direct income transfers are far more effective than adjustments to the settlement pattern'.

The problem with most urbanization policies is that they were formulated during the 1960s and 1970s when the prevailing sentiment reflected a pessimism regarding the presence of 'trickle-down' effects, which had been expected to spread the benefits of economic growth. Economists, town and regional planners, and many others set out to devise policies that would redress inequality and meet basic needs more directly. The reason for the pessimism is less clear nowadays, as many hold that, after all, the example of the newly industrializing countries shows that poverty is more rapidly relieved by a prior focus on growth and efficiency than on redistribution. Likewise, Moll (1986b) has reported that South American countries that pursued redistribution and disregarded the 'economic fundamentals' achieved less for the poor than other countries that were more concerned with those fundamentals.

As the enthusiasm for basic needs diminished so, too, did the urgency for urbanization policies, whose record has caused many to become cynical regarding government attempts to reach the poor directly. Projects were seldom really targeted at the poor and, even when they were, project benefits were usually siphoned off by middle- and upper-income groups. Further, the costs of the projects ensured that their number and scale would never meet the need.

At the same time that this reorientation was under way, planners came to realize that there were inherent constraints to the success of their schemes. For instance, an industrialization policy based on import substitution will concentrate industrial growth in the larger centres, where the market is, and industrial decentralization strategies can achieve little in the face of this except, perhaps, to induce economic inefficiency. When regional planners tried to address unemployment and poverty in peripheral regions, their most prominent effect may have been to reduce the overall rate of job creation – this certainly seems true in South Africa. Similarly, experience has shown that site-and-service and squatter-upgrade schemes may cause the poor to leave the area, as they may not be able to afford either to participate in the scheme or to pay the higher rents that the improvements enable landowners to charge.

A re-think is under way and past urbanization policies that professed to, and sometimes genuinely tried to, reach the poor directly are in the process of being replaced by a focus on management and indirect policies. This new breed of policies, through legislation, highly selective public investment and carefully located subsidies, is attempting to influence the behaviour of the market so as to deliver the needed good or service more widely. The World Bank points to

> a transition from a first generation of urban projects that stressed the development of particular housing sites to operations that are designed to have city-wide or market-wide effects. While the shift is gradual, a growing number of Bank projects are making a transition from delivering houses directly through pilot projects, to improving housing delivery systems. (*Urban Edge*, 1986, 10, 8: 1)

The policies considered in this book fall somewhat uneasily into what appears to represent a half-way stage in the reformulation of urbanization policy. One abandons growth centre policy, but then calls for secondary cities, whose record of success remains to be demonstrated. One criticizes site-and-service schemes, but then calls for a land development corporation because of the paucity of alternatives and an unwillingness to abandon the poor to the market. And so on. This fence-sitting is, however, probably all to the good since, if the policies constitute an attractive target, they are more likely to encourage debate regarding the options available to a democratic government. In this light, the more barbs they gather, the more successful this book is.

It is particularly important that one does not adopt the view of an influential member of the African National Congress that policy emerges from the political economic struggle and that prior discussion, of the sort undertaken here, is fatuous or even presumptuous. If this book has steered the debate towards greater recognition of international precedent and away from too great an obeisance to the political process or from lazy assumptions that South Africa is unique, then it will have been doubly successful.

Housing is probably the issue that springs most quickly to mind when urbanization policy is considered. A review of international experience of housing policy has showed that less developed or middle-income countries simply cannot provide housing for all their people. When governments build housing, they do it at too high a standard and make it available below price (or with a subsidized rental) to favoured constituencies, commonly civil servants. Likewise, the poor cannot afford to provide their own housing. For them the issue is one of access to rented rooms or land – hopefully, serviced land. In South Africa, even financial arrangements that take the incomes of the entire household, and not just the head of the household, into account exclude the large majority of the country's black population. The conclusion must be that, for the urban poor, the government's primary responsibility lies in increasing the supply of serviced land. The PWV, it was shown, is rather special in this respect in that, in contrast to many Third World cities, there is no shortage of land.

The resulting focus of Chapter 4 was to consider alternative ways to increase the supply of land. I looked at taxes and other policies intended to affect the market in land and, remaining sceptical about whether an improved market in land would benefit the poor, I also looked at a land development corporation, which is intended to directly improve the supply of land for low-income households.

The reasons for suggesting taxes on land are:

- to prompt owners of land that is currently un- or under-utilized either to sell or to develop that land;
- to restrain or lower prices and inhibit speculation; and,
- since one of the major reasons for high land prices is the shortage of serviced land available for development, to recover for the public sector the increases in value of land that result from public investment, with a view to enabling the public sector to sustain a high rate of investment in land.

Taxation proved to be a rather complex subject. For example, a capital gains tax, which reduces the after-tax profits that can be expected when the land is sold, would appear to be an obvious means of reducing the investment value of land and of inhibiting speculation. Since potential buyers realize that they, too, will be taxed on future price appreciation, they are expected to offer less for the land. The prediction is that the value of the land should be reduced for both existing and future landowners, culminating in lower prices. However, quite the opposite may occur since, if one expects a lower return from selling the land, then one might put off the sale. So the actual result might be a reduction in the supply of land to the market and higher prices!

Another often unforeseen outcome is that there is little reason, once the tax has taken effect, for there to be a slow-down in the rate of price increase. Tax changes are capitalized and reflected in the value of the land. Thereafter, changes in the value of the land are driven by the demand for land. The attractiveness of investing in land depends on price changes relative to the return available on other possible investments. From the point of view of the speculator, there is not now less reason to consider investing in land.

The example demonstrates that public responses to land policy should be approached with considerable caution. In a sense, it is like reading a murder mystery – one can rarely be sure how motivations, actions and outcomes are linked; a variety of unforeseen behaviours are likely.

Two taxes were viewed more favourably. One is a progressive, as opposed to the present uniform, site value tax, which would redistribute the burden of providing and extending municipal services onto wealthier groups and also promote smaller land holdings and an increase in density. The other tax is a vacant land tax, which would be implemented alongside the site value tax. Unlike most other taxes, which are politically unpopular since they affect all landowners, vacant land taxes are more discriminating and do not involve higher taxes and a possible decrease in property prices for all. Vacant land taxes enable the government to influence where development occurs and where higher densities arise.

The taxes cause landowners and developers to acquire a certain predisposition in respect of the location and form of development. While this might contribute to more desirable market outcomes, it was not clear that this would significantly improve the access of the urban poor to residential land. It was for this reason that I also suggested a land development corporation. Being aware of the problems experienced in respect of such corporations, I went to some lengths to suggest a corporation whose constitution and operating procedures would mean that it really should successfully deliver to the poor. The corporation would not be independent and, not expecting that the poor would constitute a significant constituency at the national level, I suggested that it should work for the local government. The ANC supports strong local government and it would be hoped that the local government would be mindful of the needs of the poor. Failing democratic successes for the poor at the local level, I attempted another safeguard, namely, that the major source of finance be a central development bank whose conditions for assistance would include the participation of the poor in the design and implementation of projects. This is, at any rate, necessary if cost recovery is to be successful.

Regional policy in South Africa, incorporating the homeland system, influx control, migrant labour and industrial decentralization, has represented a devastating interference in the natural processes of urbanization. Supposed explanations for the policy, such as that South Africa's cities are, or threaten to become, too large, are not worth serious consideration. In general, 'we really do not know when a city is too large or too congested, rather than poorly organized or managed. Furthermore, we have as yet learned little more than the rudiments of how to convert a settlement into a growth center and how to radiate so-called growth impulses from such centers to surrounding hinterlands . . .' (Rodwin and Sanyal 1987: 16).

The regional planner's limited policy repertoire does not, however, mean that a democratic government will be able to avoid intervention in regional economies. Political tensions are likely to give rise to attempts to redress regional inequality. Having assumed that the post-apartheid state would not control population movement and having argued that it is unable to direct the location of private investment, the appropriate policy response appears to consist largely of a secondary city strategy. In particular, the government should strive to avoid the mistakes made with the previous decentralization policy.

The latter policy appears to have been undertaken in ignorance of empirical trends. Bell (1983, 1984) has argued that labour-intensive industries were driven by competition to seek out cheaper labour in the South African periphery. I questioned this position because of the apparently lower productivity of peripheral labour, but did not deny that, from the industrialist's point of view, labour at isolated locations frequently has the marked attribute of not being unionized.

The most important point that followed from trying to explain the location trends was that industry is not free to locate where it chooses. When competition exists, an inefficient location either threatens the industry's survival or will inevitably ensure that it has to be subsidized. This is evident in the fact that about 60 per cent of the jobs benefiting from concessions would not survive without those concessions. It was also especially disillusioning to read that a decentralized job occurs at the cost of about four jobs in the cities (Wellings and Black 1986). The government's review of the programme in 1988 resulted from its fiscal constraints coupled with belated recognition of the international re-evaluation of growth centre policy and the move towards a strategy of deconcentration points and secondary cities. In general, through, one does not gain much confidence from the literature that a country can significantly affect the location of industry and employment without, at the same time, introducing economic inefficiency and retarding growth. Secondary cities can never be seen as anything more than a modest attempt to interfere in the system – while accepting that some inefficiency might result but arguing that there are specific benefits that justify this cost.

The rural development chapter (Chapter 6) was exciting for me to put together because of the way in which, initially, theory and practice came together to justify a policy of land redistribution that would contribute to both increased output and reduced inequality. The theoretical explanations were based on the different factor costs facing large and small farmers: for small farmers, land and capital are expensive and labour is cheap, whereas the reverse is true for large farmers. The result, found in many countries around the world, is that small farmers use their land intensively and employ capital very productively. Large farmers cultivate a smaller proportion of their land, are less dexterous in their use of capital and economize in the use of labour. From the point of view of urbanization policy, large farms are not conducive to the desired result – namely, many employment opportunities and higher density settlement on the land.

Unhappily, though, the theoretical predictions were not borne out because South African agriculture, unlike the agriculture of Third World countries, exhibits scale economies. In this respect, since South Africa is a food surplus country, it is not the level of output but the income generated from farming that is important. Employing this criterion, the optimum farm size considerably exceeded the size necessary for a viable small farm. Moreover, models of viable small farms that produce a household income of R6,000–7,000 per annum required so much land that it seems that the redistribution of farms would not promise to settle more people on the land. Consequently, instead of proposing the redistribution of white farms, I settled with suggestions urging that the government adopt legislation that improves conditions on the farms and favours the organization of farm labour. However, both these reforms would cause farmers to reduce their labour force and so, what at the outset constituted a hopeful means of restraining urbanization, ended up by exacerbating the move to the towns.

Lastly, the Transkeian study sought to identify the appropriate public sector response to the pattern of urbanization that is emerging in the homelands. The central question that emerged was whether migrant labourers will eventually emigrate. Graaff (1986) believes that better-off migrants will often invest in agricultural and other resources in the rural areas in order to continue to live there. In his view, therefore, many people with rural resources are unlikely to view urbanization positively and emigration will not be especially pronounced.

I find it difficult to go along with this argument. While Graaff might, to a degree, be correct at this stage, he offers little guidance for the future. First, if migrants previously had been allowed to invest in urban areas, who is to say that urban investments (especially housing) rather than rural investments would not have been preferred? Secondly, since the mining houses are heading towards a greater proportion of locally settled labour, the option of continuing to migrate will progressively diminish. My understanding of present urbanization policy is policy is that it is intended to keep out only the unemployed, not those with jobs. I therefore expect that the movement of employed Transkeians, with their families, into urban areas elsewhere in South Africa will eventually become an inexorable trend. It is the poor and the unemployed who, as a result of the massive unemployment throughout South Africa and also because of the high cost of land and services in South Africa, will, to some degree, remain in the Transkei. The question one faces is where this group will settle in the Transkei.

Given the lack of development in the towns, and the limited resources available to them, it appears that the Transkeian government intends to avoid urbanization in the homeland. A number of commentators have indicated that the government wants the poor and the unemployed to settle in the rural areas. This is less and less possible and I have provided statistics that point to the extremely rapid growth of the peri-urban areas and also of some rural villages. This growth might slow down if migrants, instead of depositing their families in such areas, take them with them to wherever they work – otherwise, there is no apparent reason for a decline in the growth rate. Landlessness is already widespread and for many of those who have some land the cost of farming appears to be prohibitive. The Transkei government's rural development strategy hinges on the creation of a class of master farmers, in which case two-thirds of the people on the land will have to relocate. In that their aim is to keep the unemployed out of the towns, this is a bit contradictory – or perhaps the government is unwilling to face the urban consequences that would result. None the less, it is difficult to imagine an alternative to the creation of this class of farmers if agricultural development is to be pursued. Therefore, even when a democratic government comes to power, the rural development strategy will probably differ only in so far as there will be more resources devoted to small farmers, extension services, technology and access to markets.

In the Transkei, the major urbanization problem, which is accentuated by rural development and the relocation of population, is that the economic base of the towns is barely growing and, as there is no reason to expect any economic expansion, the towns are on the way to becoming agglomerations of desperately poor people. The Transkeian government and, in the future, regional and central government departments should envisage intense pressure on their ability to provide services for the people. This, in turn, has raised the issue of where given levels of investments should be undertaken. Governments normally organize their investment programmes around some preconception of an urban hierarchy but, in the Transkei, some towns are growing while others are declining. The hierarchy is in a state of flux, which is a natural response to the reorganization and relocation of administrative functions, as the economic means of the inhabitants change in relation to the exit of formerly migrant labour and as, hopefully, rural development proceeds.

In the context of rapid and unpredictable change, I suggest that, rather than undertake planning studies and prepare social investment plans based on a targeted urban hierarchy, it would be a lot simpler and cheaper simply to conduct periodic aerial surveys and to liaise with the communities identified. These people will, largely, be living outside designated urban areas and the need for community organization arises because of the lack of an effective local government. Consequently, the settlements will usually fall under a headman or chief who has neither the resources nor the ability to undertake the necessary administration and investment. In other words, my proposal consists of identifying areas of population concentration, facilitating community organization and working with the communities to undertake the investment programme mutually agreed to. This programme would obviously be objected to by the 'traditional' headmen or chiefs, since it would entail a loss of power and prestige, but it is difficult to imagine that an ANC-led government would prolong their positions of authority over and against democratic outcomes.

Owing to the widespread interest in the ANC, there is one other sense in which I believe the book provides a service. During the period when this book was being written overseas, I contacted the ANC's United Nations mission with a view to ascertaining whether the policies considered in this book were in the 'ball park'. The response was that 'We have no policies in this area. We have still to consider the issues'. The same view was repeated to me in London and again in South Africa by observers of the ANC. Although I think this vacuum was understandable in the past, it is less so now. In my view, there are many whites in South Africa who can see through the government's propaganda against the ANC and whose wavering commitment to a free and democratic society would be strengthened by clarity regarding what it is that the ANC wants. Nevertheless, the difficulties in drawing up positions of this sort are acute, given that the ANC is an alliance, not an individual political party, so the ANC is to be praised for proposing a constitutional outline. My hope is that this book can contribute to progress in the formulation of urbanization policy.

In this book I have attempted to combine poor data and an optimistic sense of the future with urbanization theory and practice from elsewhere in the world. The outcome has been a lengthy, at times complex, treatise on the types of issues a democratic government will confront and suggestions regarding the nature of appropriate policies. It has been written in the hope that it can contribute to the evolution of a free and democratic South Africa.

Appendix
The Freedom Charter and the ANC's Constitutional Guidelines

The Freedom Charter

WE THE PEOPLE OF SOUTH AFRICA, DECLARE FOR ALL
OUR COUNTRY AND THE WORLD TO KNOW:

That South Africa belongs to all who live in it, black and white,
and that no government can justly claim authority unless it is based
on the will of the people;
that our people have been robbed of their birthright to land,
liberty and peace by a form of government founded on injustice
and inequality; that our country will never be prosperous or free
until all our people live in brotherhood, enjoying equal rights and
opportunities;
that only a democratic state, based on the will of all the people, can
secure to all their birthright, without distinction of colour, race,
sex or belief;
And therefore, we, the people of South Africa, black and white
together – equals, countrymen and brothers – adopt this Freedom
Charter. And we pledge ourselves to strive together, sparing
neither strength nor courage, until the democratic changes here
set out have been won.

1:    THE PEOPLE SHALL GOVERN!
1.1   Every man and woman shall have the right to vote for and
      to stand as a candidate for all bodies which make laws;
1.2   All people shall be entitled to take part in the administration
      of the country;

1.3 The rights of the people shall be the same regardless of sex, race and colour;

1.4 All bodies of minority rule, advisory boards, councils and authorities, shall be replaced by democratic organs of self-government.

2: ALL NATIONAL GROUPS SHALL HAVE EQUAL RIGHTS!

2.1 There shall be equal status in the bodies of state, in the courts and in the schools, for all national groups and races;

2.2 All the people shall have equal right to use their own languages and to develop their own folk cultures and customs;

2.3 All national groups shall be protected by law against insults to their race and national pride;

2.4 The preaching and practice of national, race or colour discrimination and contempt shall be a punishable crime;

2.5 All apartheid laws and practices shall be set aside.

3: THE PEOPLE SHALL SHARE IN THE COUNTRY'S WEALTH!

3.1 The national wealth of our country, the heritage of all South Africans, shall be restored to the people;

3.2 The mineral wealth beneath the soil, the Banks and the monopoly industry shall be transferred to the ownership of the people as a whole;

3.3 All other industry and trade shall be controlled to assist the well-being of the people;

3.4 All people shall have equal rights to trade where they choose, to manufacture and to enter all trades, crafts and professions.

4: THE LAND SHALL BE SHARED AMONG THOSE WHO WORK IT!

4.1 Restrictions of land ownership on a racial basis shall be ended, and all the land redivided amongst those who work it to banish famine and land hunger;

4.2 The state shall help the peasants with implements, seed, tractors and dams to save the soil and assist the tillers;

4.3 Freedom of movement shall be guaranteed to all who work on the land;

4.4 All people shall have the right to occupy land wherever they choose;

4.5 People shall not be robbed of their cattle, and forced labour and farm prisons shall be abolished.

5:     ALL SHALL BE EQUAL BEFORE THE LAW!

5.1   No one shall be imprisoned, deported or restricted without a fair trial;

5.2   No one shall be condemned by the order of any government official;

5.3   The courts shall be representative of all the people;

5.4   Imprisonment shall only be for serious crimes against the people, and shall aim at re-education not vengeance;

5.5   The police force and army shall be open to all on an equal basis, they shall be the helpers and protectors of the people;

5.6   All laws which discriminate on the grounds of race, colour or belief shall be repealed.

6:     ALL SHALL ENJOY EQUAL HUMAN RIGHTS!

6.1   The law shall guarantee to all their right to speak, to organise, to meet together, to publish, to preach, to worship and to educate their children;

6.2   The privacy of the house from police raids shall be protected by law;

6.3   All shall be free to travel without restriction from country-side to town, from province to province, and from South Africa abroad;

6.4   Pass laws, permits and all other laws restricting these freedoms shall be abolished.

7:     THERE SHALL BE WORK AND SECURITY!

7.1   All who work shall be free to form trade unions, to elect their officers and to make wage agreements with their employers;

7.2   The state shall recognise the right and duty of all to work, and to draw full unemployment benefits;

7.3   Men and women of all races shall receive equal pay for equal work;

7.4   There shall be a forty hour working week, a national minimum wage, paid annual leave, and sick leave for all workers, and maternity leave on full pay for all working mothers;

7.5   Miners, domestic workers, farm workers and civil servants shall have the same rights as all others who work;

7.6   Child labour, compound labour, the tot system and contract labour shall be abolished.

8:     THE DOORS OF LEARNING AND OF CULTURE SHALL BE OPENED!

8.1   The government shall discover, develop and encourage national talent for the enhancement of our cultural life;

8.2   All cultural treasures of mankind shall be open to all, by free exchange of books, ideas and contact with other lands;

8.3   The aim of education shall be to teach the youth to love their people and their culture, to honour human brotherhood, liberty and peace;

8.4   Education shall be free, compulsory, universal and equal for all children;

8.5   Higher education and technical training shall be opened to all by means of state allowances and scholarships awarded on the basis of merit;

8.6   Adult illiteracy shall be ended by a mass state education plan;

8.7   Teachers shall have all the rights of other citizens;

8.8   The colour bar in cultural life, in sports and in education shall be abolished.

9:   THERE SHALL BE HOUSES, SECURITY AND COMFORT!

9.1   All people shall have the right to live where they choose, to be decently housed, and to bring up their families in comfort and security;

9.2   Unused housing space to be made available to the people;

9.3   Rent and prices shall be lowered, food plentiful and no one shall go hungry;

9.4   A preventative health scheme shall be run by the state;

9.5   Free medical care and hospitalisation shall be provided for all, with special care for mothers and young children;

9.6   Slums shall be demolished, and new suburbs built where all have transport, roads, lighting, playing fields, creches and social centres;

9.7   The aged, the orphans, the disabled and the sick shall be cared for by the state;

9.8   Rest, leisure and recreation shall be the right of all;

9.9   Fenced locations and ghettos shall be abolished, and laws which break up families shall be repealed.

10:   THERE SHALL BE PEACE AND FRIENDSHIP!

10.1   South Africa shall be a fully independent state, which respects the rights and sovereignty of all nations;

10.2   South Africa shall strive to maintain world peace and the settlement of all international disputes by negotiation, not war;

10.3   Peace and friendship amongst all our people shall be secured by upholding the equal rights, opportunities and status of all;

10.4 The people of the protectorates – Basutoland, Bechuanaland and Swaziland – shall be free to decide for themselves their own future;

10.5 The rights of all the people of Africa to independence and self-government shall be recognised, and shall be the basis of close co-operation.

Let all who love their people and their country now say, as we say here:

"THESE FREEDOMS WE WILL FIGHT FOR, SIDE BY SIDE, THROUGHOUT OUR LIVES, UNTIL WE HAVE WON OUR LIBERTY."

The ANC's Constitutional Guidelines

THE STATE

(A) South Africa shall be an independent, unitary, democratic and non-racial state.

(B) Sovereignty shall belong to the people as a whole and shall be exercised through one central legislature, executive, judiciary and administration. Provision shall be made for the deregulation of the powers of the central authority to subordinate administrative units for purposes of more efficient administration and democratic participation.

(C) The institution of hereditary rulers and chiefs shall be transformed to serve the interests of the people as a whole in conformity with the democratic principles embodied in the constitution.

(D) All organs of government, including justice, security and armed forces, shall be representative of the people as a whole, democratic in their structure and functioning, and dedicated to defending the principles of the constitution.

FRANCHISE

(E) In the exercise of their sovereignty, the people shall have the right to vote under a system of universal suffrage based on the principle of one person/one vote.

(F) Every voter shall have the right to stand for election and to be elected to all national bodies.

NATIONAL IDENTITY

(G) It shall be state policy to promote the growth of a single national identity and loyalty binding on all South Africans. At the same time, the state shall recognise the linguistic and cultural diversity of the people and provide facilities for free linguistic and cultural development.

BILL OF RIGHTS AND AFFIRMATIVE ACTION

(H) The Constitution shall include a Bill of Rights based on the Freedom Charter. Such a Bill of Rights shall guarantee the fundamental human rights of all citizens, irrespective of race, colour, sex or creed, and shall provide appropriate mechanisms for their protection and enforcement.

(I) The state and all social institutions shall be under a constitutional duty to eradicate race discrimination in all its forms.

(J) The state and all social institutions shall be under a constitutional duty to take active steps to eradicate, speedily, the economic and social inequalities produced by racial discrimination.

(K) The advocacy or practice of racism, fascism, nazism or the excitement of ethnic or regional exclusiveness or hatred shall be outlawed.

(L) Subject to clauses (I) and (K) above, the democratic state shall guarantee the basic rights and freedoms, such as freedom of association, thought, worship and the press. Furthermore, the state shall have the duty to protect the right to work and guarantee the right to education and social security.

(M) All parties which conform to the provisions of (I) to (K) above shall have the legal right to exist and to take part in the political life of the country.

ECONOMY

(N) The state shall ensure that the entire economy serves the interests and well-being of the entire population.

(O) The state shall have the right to determine the general context in which economic life takes place and define and limit the rights and obligations attaching to the ownership and use of productive capacity.

(P) The private sector of the economy shall be obliged to cooperate with the state in realising the objectives of the Freedom Charter in promoting social well-being.

(Q) The economy shall be a mixed one, with a public sector, a private sector, a co-operative sector and a small-scale family sector.

(R) Co-operative forms of economic enterprise, village industries and small-scale family activities shall be supported by the state.

(S) The state shall promote the acquisition of management, technical and scientific skills among all sections of the population, especially the blacks.

(T) Property for personal use and consumption shall be constitutionally protected.

LAND

(U) The state shall devise and implement a land reform programme that will include and address the following issues: Abolition of racial restrictions on ownership and use of land, implementation of land reform in conformity with the principles of affirmative action, taking into account the status of victims of forced removals.

WORKERS

(V) A charter protecting workers' trade union rights, especially the right to strike and collective bargaining, shall be incorporated into the constitution.

WOMEN

(W) Women shall have equal rights in all spheres of public and private life and the state shall take affirmative action to eliminate inequalities and discrimination between the sexes.

THE FAMILY

(X) The family, parenthood and children's rights shall be protected.

INTERNATIONAL

(Y) South Africa shall be a non-aligned state committed to the principles of the Charter of the OAU and the Charter of the UN and to the achievement of national liberation, world peace and disarmament.

Bibliography

Abedian, I. (1983) 'A quantitative review of the economy of Transkei', *South African Journal of Economics*, 51, 2: 252–69.

Acharya, B. B. (1987) 'Policy of land acquisition and development: analysis of Indian experience', *Third World Planning Review*, 9, 2: 99–116.

Adam, H. (1971) *Modernizing Racial Domination*, University of California Press.

Adam, H. (1983) 'Outside influence on South Africa: Afrikanerdom in disarray', *Journal of Modern African Studies*, 21, 2: 235–51.

Adam, H. (1987) 'Implications of the May 1987 white elections in South Africa', Paper presented at a seminar of the Centre for African Studies, University of Cape Town, May 20.

Adam, H. (1988) 'Exile and resistance: the African National Congress, the South African Communist Party and the Pan Africanist Congress', in P. Berger and B. Godsell (eds), *A Future South Africa: Visions, Strategies and Realities*, Tafelberg: Human and Rousseau, pp. 93–124.

Adam, H. and Moodley, K. (1986) *South Africa without Apartheid: Dismantling Racial Domination*, University of California Press.

Addleson, M. and Tomlinson, R. (1985) 'Decentralisation', *Leadership SA: Human Resources 1985–86* (special edition): 40–4.

Addleson, M., Pretorius, F. and Tomlinson, R. (1985) 'The impact of industrial decentralisation policy: the businessman's view', *South African Geographical Journal*, 67, 2: 39–60.

Addleson, M., Pretorius, F. and Tomlinson, R. (1989) *Industrial Trends and Prospects in Natal/KwaZulu*, Pietermaritzburg: Natal Town and Regional Planning Commission.

Archer, R. W. (1974) 'The leasehold system of urban development: land tenure, decision-making and the land market in urban development and land use', *Regional Studies*, 8: 225–38.

Archer, S. (1986) 'Economic ends and political means in the Freedom Charter: problems in interpretation and assessment', Department of Economics, University of Cape Town (mimeo).

Ardington, E. M. (1984) 'Decentralised industry, poverty and development in rural KwaZulu', *Working Paper No. 10*, Development Studies Unit, University of Natal, Durban.

Bahl, R. W. (1979) 'The practice of urban property taxation in less developed countries', in R. W. Bahl (ed.), *The Taxation of Urban*

Property in Less Developed Countries, University of Wisconsin Press, pp. 9–50.

Baran, P. A. and Sweezy, P. M. (1966) *Monopoly Capital*, New York: Monthly Review Press.

Baross, P. (1983) 'The articulation of land supply for popular settlements in Third World cities', in S. Angel, R. W. Archer, S. Tanphiphat and E. A. Wegelin (eds), *Land for Housing the Poor*, Thailand: Select Books, pp. 180–210.

Baskin, J. (1984) 'Access to land in the Transkei', Paper delivered at the Second Carnegie Inquiry into Poverty and Development in Southern Africa, University of Cape Town, 13–19 April.

Baum, W. C. and Tolbert, S. M. (1985) *Investing in Development: Lessons of World Bank Experience*, Oxford University Press, published for the World Bank.

Beckett, D. (1986) *Permanent Peace*, Braamfontein: Saga Press.

Bell, R. T. (1983) 'The growth and structure of manufacturing employment in Natal', *Occasional Paper No. 7*, Institute of Social and Economic Research, University of Durban–Westville.

Bell, R. T. (1984) 'The state, the market and the interregional distribution of industry in South Africa', Paper delivered at the Second Carnegie Inquiry into Poverty and Development in Southern Africa, University of Cape Town, 13–19 April.

Bell, R. T. (1985) 'Issues in South African unemployment', *South African Journal of Economics*, 53, 1: 24–38.

Bell, R. T. (1986) 'The role of regional policy in South Africa', *Journal of Southern African Studies*, 12, 2: 276–92.

Bell, R. T. (1987a) 'International competition and industrial decentralisation in South Africa', *World Development*, 15, 10/11: 1291–307.

Bell, R. T. (1987b) 'Rethinking industrial decentralisation', in R. Tomlinson and M. Addleson (eds), *Regional Restructuring under Apartheid: Contemporary Perspectives on Urban and Regional Policies in South Africa*, Johannesburg: Ravan Press, 1987, pp. 207–21.

Berry, B. J. L. (1972) 'Hierarchical diffusion: The basis of developmental filtering and spread in a system of growth centres', in P. W. English and R. C. Mayfield (eds), *Man, Space and Environment*, London: Oxford University Press, pp. 340–60.

Berry, R. A. and Cline, W. R. (1979) *Agrarian Structure and Productivity in Developing Countries*, Baltimore: Johns Hopkins University Press.

Bienen, H. (1984) 'Urbanisation and Third World stability', *World Development*, 12, 7: 661–91.

Blumenfeld, J. (1986) 'Investment, savings and the capital market in South Africa', Paper presented at the conference 'The Southern African Economy after Apartheid', Centre for Southern African Studies, University of York, 29 September – 2 October.

Boden, R. (1987) 'Urban development – an octopus?' Paper presented at a workshop of the Development Society, Johannesburg.

Boleat, M. (1987) 'Housing finance institutions', in L. Rodwin (ed.), *Shelter, Settlement, and Development*, Boston: Allen & Unwin, pp. 151–78.

Boulle, L. (1984) *Constitutional Reform and the Apartheid State*, New York: St Martin's Press.

Brennan, E. M. and Richardson, H. W. (1986) 'Urbanisation and urban policy in Sub-Saharan Africa', *African Urban Quarterly*, 1, 1: 20–42.

Bundy, C. (1972) 'The emergence and decline of a South African peasantry', *African Affairs*, 71: 369–88.

Bureau for Market Research (1981) *Income and Expenditure Patterns of Black Households in Transkei, 1979*, Pretoria: University of South Africa.

Bureau of Market Research (1984) *Income and Expenditure Patterns of Households in KwaNdebele*, Pretoria: University of South Africa.

Burgess, R. (1982) 'Self-help housing advocacy: a curious form of radicalism. A critique of the work of John F. C. Turner', in P. M. Ward (ed.), *Self-Help Housing: A Critique*, London: Mansell Publishing, pp. 55–97.

Butcher, C. (1985) 'Planning for rural development: A political–economic study of agricultural policy in Zimbabwe', Masters thesis completed in the Department of Town and Regional Planning, University of the Witwatersrand, Johannesburg.

Butcher, C. (1986) 'Low income housing in Zimbabwe: A case study of the Epworth squatter upgrading programme', *RUP Occasional Paper No 6*, Harare: University of Zimbabwe.

Chambers, R. (1980) 'Rural poverty unperceived: problems and remedies', Washington DC: World Bank, *Staff Working Paper No. 400*.

Charney, C. (1984) 'Class conflict and the National Party split', *Journal of Southern African Studies*. 10, April: 269–82.

Cilliers, S. P. (1986) 'Demise of the Dompas: from influx control to orderly urbanisation', *Indicator SA*, 4 (January): 95–8.

Cline, W. R. (1977) 'Policy instruments for rural income redistribution', in C. R. Frank and R. C. Webb (eds), *Income Distribution and Growth in the Less-Developed Countries*, Washington DC: The Brookings Institution, pp. 281–336.

Cobbett, M. (1986) 'The land question in South Africa: A preliminary assessment', Paper presented at the conference on 'The Southern African Economy after Apartheid', Centre for Southern African Studies, University of York, 29 September – 2 October.

Cobbett, W. (1986) 'A test case for "planned urbanisation"', *Work in Progress*, 42: 25–30.

Cobbett, W. (1987) 'Onverwacht and the emergence of a regional labour market in South Africa', in R. Tomlinson and M. Addleson (eds), *Regional Restructuring under Apartheid: Contemporary Perspectives on Urban and Regional Policies in South Africa*, Johannesburg: Ravan Press, pp. 241–52.

Cobbett, W., Glaser, D., Hindson, D. and Swilling, M. (1985) 'Regionalisation, federalism and the reconstruction of the South African State', *South African Labour Bulletin*, 10, 5: 87–116.

Cobbett, W., Glaser, D., Hindson, D. and Swilling, M. (1987) 'South Africa's regional political economy: a critical analysis of reform strategy in the 1980s', in R. Tomlinson and M. Addleson (eds), *Regional Restructuring under Apartheid: Contemporary Perspectives on Urban and*

Regional Policies, Johannesburg: Ravan Press, pp. 1–27.

Collier, P. and Lal, D. (1980) 'Poverty and growth in Kenya', Washington DC: World Bank, *Staff Working Paper No. 389.*

Connolly, P. (1982) 'Uncontrolled settlements and self-build: what kind of solution?' in P. M. Ward (ed.), *Self-Help Housing: A Critique*, London: Mansell Publishing, pp. 141–74.

Cooper, D. (1986) 'Ownership and control of agriculture in South Africa', Paper presented at the conference on 'The Southern African Economy after Apartheid', Centre for Southern African Studies, University of York, 29 September – 2 October.

Davies, Bristow, Small and Associates (1986) *Towns of South Africa. Special report: Black Towns of the PWV* (consultants' report).

Davies, R. (1981) 'The spatial formation of the South African city', *GeoJournal*, supplementary issue, 2: 59–72.

Davies, R. (1986) 'Nationalisation, socialisation and the Freedom Charter', Paper presented at the conference on 'The Southern African Economy after Apartheid', Centre for Southern African Studies, University of York, 29 September – 2 October.

De Lange, A. R. (1984) 'Demographic tendencies, technological development and the future of apartheid', Paper presented at the conference on 'Economic Development and Racial Domination', University of the Western Cape, 8–10 October.

Democratic Party (1989) 'A government in the making' (pamphlet).

Development Bank of Southern Africa (1987) *Transkei Development File*, Sandton.

Devereux, S. (1983) 'South African income distribution, 1900–1980', University of Cape Town, *Saldru Working Paper No. 51.*

De Vos, T. J. (1986) 'Financing low-cost housing', Paper presented at a seminar on 'Finance for Low-cost Housing' organized by the South African Institute of Building. Pretoria: Council for Scientific and Economic Research, August.

De Vos, T. J. (1987) 'The complexity of housing in South Africa', Paper presented at the 2nd Techno-Economics Symposium, Pretoria: Council for Scientific and Industrial Research, 28 January.

De Vos, T. J. (n.d.) 'Regional housing requirements and affordability in South Africa', Pretoria: National Building Research Institute (mimeo).

Dewar, D. (1986) 'Some central issues of urban development', Paper presented at the conference on 'Directions for the Growth of the Metropolitan Region: The Second Century – Johannesburg', Rand Afrikaans University, 16 and 17 October.

Dewar, D., Todes, A. and Watson, V. (1984) 'Issues of regional development in peripheral regions of South Africa, with particular reference to settlement policy: The case of the Transkei', *Working Paper No. 29*, Urban Problems Research Unit, University of Cape Town.

Dewar, D., Todes, A. and Watson, V. (1986) *Regional Development and Settlement Policy: Premises and Prospects*, London: Allen & Unwin.

De Wet, C. J. (1987) 'Land tenure and rural development: some issues relating to the Transkei/Ciskei region', *Development Southern Africa*, 4,

3: 459–78.

Dicken, P. (1986) *Global Shift: Industrial Change in a Turbulent World*, London: Harper & Row.

Doebele, W. (1979) 'Land readjustment as an alternative to taxation for the recovery of betterment: the case of South Korea', in R. W. Bahl (ed.), *The Taxation of Urban Property in Less Developed Countries*, University of Wisconsin Press, pp. 162–90.

Doebele, W. (1983) 'Concepts of urban land tenure', in H. B. Dunkerley (ed.), *Urban Land Policy: Issues and Opportunities*, Oxford University Press, published for the World Bank, pp. 63–107.

Doebele, W. (1987) 'Land policy', in L. Rodwin (ed.), *Shelter, Settlement, and Development*, Boston: Allen & Unwin, pp. 110–32.

Doebele, W., Grimes Jr, O. F. and Linn, J. F. (1979) 'Participation of beneficiaries in financing urban services: valorization charges in Bogota, Colombia', *Land Economics*, 55, 1: 73–92.

Dunkerley, H. B. (1983) 'Introduction and Overview', in H. B. Dunkerley (ed.), *Urban Land Policy: Issues and Opportunities*, Oxford University Press, published for the World Bank, pp. 3–39.

Ehlers, J. H. (1982) 'Pendelaars in KwaNdebele', Pretoria: Human Sciences Research Council, Navorsingsbevinding MN 93.

Eicher, C. K. (1986) 'Facing up to Africa's food crisis', in J. Ravenhill (ed.), *Africa in Economic Crisis*, New York: Columbia University Press, pp. 149–80.

Elias, C. (1984) 'A housing study: legislation and the control of the supply of urban African accommodation', Paper presented at the Second Carnegie Inquiry into Poverty and Development in Southern Africa, University of Cape Town, 13–19 April.

Ellman, M. (1981) 'Agricultural productivity under socialism', *World Development*, 9, 9/10: 979–89.

El-Shakhs, S. (1972) 'Development, primacy and systems of cities', *Journal of Developing Areas*, 7, 1: 181–206.

Engels, F. (1970) *The Housing Question*, Moscow: Progress Publishers.

Fainstein, N. and Fainstein, S. (1985) 'Is state planning necessary for capital? The US case', *International Journal of Urban and Regional Research*, 9: 485–507.

Falcoff, M. (1987) 'An authoritarian by popular consent', *New York Times Book Review*, 18 January.

Friedman, J. and Weaver, C. (1979) *Territory and Function*, London: Edward Arnold.

Friedman, S. (1988) *Reform Revisited*, Johannesburg: South African Institute of Race Relations.

Freund, W. (1984) 'Forced resettlement and the political-economy of South Africa', *Review of African Political Economy*, 29: 49–76.

Freund, W. (1986a) 'Some unasked questions in politics: South African slogans and debates', *Transformation*, 1: 118–37.

Freund, W. (1986b) 'South African business ideology, the crisis and the problems of redistribution', Paper presented at the conference on

'The Southern African Economy after Apartheid', Centre for Southern African Studies, University of of York, 29 September – 2 October.

Galbraith, J. K. (1979) *The Nature of Mass Poverty*, Cambridge, Mass.: Harvard University Press.
Gilbert, A. (1982a) 'The housing of the urban poor', in A. Gilbert and J. Gugler (eds), *Cities, Poverty and Development: Urbanization in the Third World*, Oxford University Press, pp. 81–115.
Gilbert, A. (1982b) 'Urban and regional systems: a suitable case for treatment?' in A. Gilbert and J. Gugler (eds), *Cities, Poverty and Development: Urbanization in the Third World*, Oxford University Press, pp. 162–97.
Gilbert, A. and Gugler, J. (1982) *Cities, Poverty and Development: Urbanization in the Third World*, Oxford University Press.
Giliomee, H. and Schlemmer, L. (1985) *Up against the Fences: Poverty, Passes and Privilege in South Africa*, Cape Town: David Philip.
Glaser, D. (1988) 'Democracy, socialism, and the future', *Work in Progress*, 56/57: 28–30.
Gore, C. (1984) *Regions in Question: Space, Development Theory and Regional Policy*, London: Methuen.
Gottschalk, K. (1977) 'Industrial decentralisation, jobs and wages', *South African Labour Bulletin*, 3, 5: 50–8.
Graaff, J. (1986) 'The present state of black urbanisation in the South African homelands and some future scenarios', Paper presented at the biennial conference of the Development Society of Southern Africa, University of Cape Town, September.
Green, R. H. and Allison, C. (1986) 'The World Bank's agenda for accelerated development: Dialectics, doubts and dialogues', in J. Ravenhill (ed.), *Africa in Economic Crisis*, New York: Columbia University Press, pp. 60–84.
Greenberg, S. B. (1980) *Race and State in Capitalist Development: South Africa in Comparative Perspective*, Johannesburg: Ravan Press.
Greenberg, S. B. (1984) 'Ideological struggles within the South African state', Paper presented at the conference on 'Economic Development and Racial Domination', University of the Western Cape, 8–10 October.
Grimes Jr, O. F. (1977) 'Urban land policy: social appropriation of betterment', in P. Downing (ed.), *Local Service Pricing Policies and their Effect on Urban Spatial Structure*, Vancouver: University of British Columbia Press, pp. 360–434.

Hall, P. (1987) 'Metropolitan settlement strategies', in L. Rodwin (ed.), *Shelter, Settlement and Development*, Boston: Allen & Unwin, pp. 236–59.
Hamilton, F. E. I. and Linge, G. J. R. (1983) 'Regional economies and industrial systems', in F. E. I. Hamilton and G. J. R. Linge (eds), *Spatial Analysis, Industry and the Industrial Environment. Vol. 3 Regional Economies and Industrial Systems*, New York: John Wiley and Sons, pp. 1–39.
Hardie, G. J., Hart, T. and Strellitz, J. (1987) 'Housing finance and

homeownership: an investigation of practices, perceptions and problems in the context of five urban townships', *Future Build*, 6: 27–32.

Hardoy, J. and Sattherwaite, D. (1981) *Shelter: Need and Response*, New York: John Wiley and Sons.

Harris, C. L. (1979) 'Land taxation in Taiwan: selected aspects', in R. W. Bahl (ed.), *The Taxation of Urban Property in Less Developed Countries*, University of Wisconsin Press, pp. 191–206.

Harris, J. R. (1972) 'A housing policy for Nairobi', in J. Hutton (ed.), *Urban Challenge in East Africa*, Nairobi: East African Publishing House, pp. 39–56.

Hattingh, H. (1986) 'Skewe inkomeverdeling in die landbou se uitdaging aan landboubeleid', Paper delivered at Lanvokon '86, 11 February.

Hawkins Associates (1980) 'The physical and spatial basis for Transkei's first five-year development plan', Report to Transkei's National Planning Commission.

Haysom, N. (1985) 'The Langa shootings and the Kannemeyer Commission of Enquiry', in *SA Review 3*, Johannesburg: Ravan Press, pp. 278–89.

Hegedus, J. and Tosics, I. (1983) 'Housing classes and housing policy: some changes in the Budapest housing market', *International Journal of Urban and Regional Research*, 7, 4: 467–94.

Hindson, D. (1987a) *Pass Controls and the Urban African Proletariat*, Johannesburg: Ravan Press.

Hindson, D. (1987b) 'Orderly urbanisation and influx control: from territorial apartheid to regional spatial ordering in South Africa', in R. Tomlinson and M. Addleson (eds), *Regional Restructuring under Apartheid: Contemporary Perspectives on Urban and Regional Policies in South Africa*, Johannesburg: Ravan Press, pp. 74–105.

Hofmeyr, J. F. (1985) *A Survey of Small Industries in Selected Areas of Transkei*, A report prepared for Transido, Economic Research Unit, University of Natal.

Hudson, P. (1986) 'The Freedom Charter and the theory of the national democratic revolution', *Transformation*, 1: 6–38.

Hunt, D. (1984) *The Impending Crisis in Kenya: The Case for Land Reform*, London: Gower.

Huzinec, G. A. (1978) 'The impact of industrial decision-making upon the Soviet urban hierarchy', *Urban Studies*, 15, 2: 139–48.

Innes, D. (1985) 'The real world of the Left', *Frontline*, August: 10–18.

International Bank for Reconstruction and Development (1975) *Housing*, Washington DC: World Bank Sector Policy Paper.

International Bank for Reconstruction and Development (1984) *Toward Sustained Development in Sub-Saharan Africa: A Joint Program for Action*, Washington DC: World Bank.

International Labour Organization (1972) *Employment, Incomes and Equality: A Strategy for Increasing Productive Employment in Kenya*, Geneva.

Jameson, K. P. and Wilber, C. K. (1981) 'Socialism and development: editor's introduction', *World Development*, 9, 9/10: 803–11.

Johnson, P. D. and Campbell, P. R. (1982) *Detailed Statistics on the Population of South Africa by Race and Urban/Rural Residence: 1950–2010*, Washington DC: International Demographic Data Center, US Bureau of the Census.

Karis, T. G. (1983/4) 'Revolution in the making: black politics in South Africa', *Foreign Affairs*, 62, 2: 378–406.
Karis, T. G. (1986/7) 'South African liberation: the Communist factor', *Foreign Affairs*, 65, 2: 267–88.
Karis, T. G. and Carter, G. M. (eds) (1977) *From Protest to Challenge. A Documentary History of African Politics in South Africa, 1882–1964. Volume 3. Challenge and Violence, 1953–64*, Stanford University: Hoover Institution Press.
Khodzhaev, D. G. and Khorev, B. S. (1978) 'The concept of a unified settlement system and the planned control of the growth of towns in the USSR', in L. S. Bourne and J. W. Simmons (eds), *Systems of Cities: Readings on Structure, Growth and Policy*, New York: Oxford University Press, pp. 511–18.
Kitay, M. G. (1985) 'Financing land acquisition', in M. G. Kitay (ed.), *Land Acquisition in Developing Countries: Policies and Procedures of the Public Sector*, Oelgeschlager, Gunn and Hain, in association with the Lincoln Institute of Land Policy, pp. 87–95.
Knight, J. B. (1986) 'A comparative analysis of South Africa as a semi-industrialised country', Paper presented at the conference on 'The Southern African Economy after Apartheid', Centre for Southern African Studies, University of York, 29 September – 2 October.
Kok, P. C. (1986) 'Population research: A summarised review and evaluation of theoretical contributions, strategies and policy instruments, with specific reference to the South African situation', *Research Finding SN-250*, Pretoria: Human Sciences Research Council.

La Grange, A. (1986) 'Access to low income housing in South Africa', Masters thesis in the Department of Town and Regional Planning, University of the Witwatersrand, Johannesburg.
Legassick, M. (1985) 'South Africa in crisis: what route to democracy?' *African Affairs*, 84, 337: 587–604.
Lele, U. (1975) *The Design of Rural Development: Lessons from Africa*, Baltimore: Johns Hopkins University Press.
Lelyveld, J. (1986) 'The night riders aboard bus no. 4174', *Weekly Mail*, 18–24 April: 12.
Lemon, A. (1987a) *Apartheid in Transition*, Aldershot: Gower.
Lemon, A. (1987b) 'Re-designing the apartheid city', Paper presented at a seminar of the Centre for African Studies, University of Cape Town, 26 August.
Le Roux, P. (1986) 'The state as economic actor: a review of the divergent perceptions of economic issues', Paper presented at the conference on 'The Southern African Economy after Apartheid', Centre for Southern African Studies, University of York, 29 September – 2 October.
Lewis, O. (1961) *The Children of Sanchez: Autobiography of a Mexican*

Family, New York: Random House.

Leys, C. (1975) *Underdevelopment in Kenya: The Political Economy of Neo-Colonialism*, Los Angeles: University of California Press.

Lijphart, A. (1985) *Power-Sharing in South Africa*, Berkeley: University of California, Institute of International Studies.

Lim, G. C. (1985) 'Korea: Land-acquisition policies and procedures of the Korean Land Development Corporation', in M. G. Kitay (ed.), *Land Acquisition in Developing Countries: Policies and Procedures of the Public Sector*, Oelgeschlager, Gunn and Hain, in association with the Lincoln Institute of Land Policy, Appendix A, pp. 131–45.

Lipton, M. (1983) 'Labour and poverty', Washington DC: World Bank, *Staff Working Paper No. 616.*

Lipton, M. (1985) *Capitalism and Apartheid. South Africa, 1910–1986*, Aldershot: Wildwood House.

Livingstone, I. (1981) *Rural Development, Employment and Incomes in Kenya*, Addis Ababa: ILO (JASPA).

Lodge, T. (1983/4) 'The African National Congress in South Africa, 1976–1983: guerilla war and armed propaganda', *Journal of Contemporary African Studies*, 3, 1/2: 153–80.

Lombard, J. A. and Du Pisanie, J. A. (1985) *Removal of Discrimination against Blacks in the Political Economy of the Republic of South Africa*, Johannesburg: Associated Chambers of Commerce.

Louw, L. and Kendall, F. (1986) *South Africa: The Solution*, Bisho: Amagi Publications.

Lozano, E. E. (1975) 'Housing the urban poor in Chile: contrasting experiences under "Christian Democracy" and "Unidad Popular"', in *Latin American Urban Research, vol. 5: Urbanization and Inequality: The Political Economy of Urban and Rural Development in Latin America*, Beverly Hills, CA: Sage Publications, pp. 180–95.

Maasdorp, G. (1982) 'Industrial decentralisation and the economic development of the Homelands', in R. Schrire (ed.), *South Africa: Public Policy Perspectives*, Cape Town: Juta, pp. 223–68.

Mabin, A. (1988) 'Households, history and black urbanisation: response to Graaff', *Development Southern Africa*, 5, 3: 292–301.

Mabin, A. (1989) 'Struggle for the city: urbanisation and political strategies of the South African state', *Social Dynamics*, 15, 1: 1–28.

McCarthy, J. J. (1986) 'Contours of capital's negotiating agenda', *Transformation*, 1: 130–7.

McCarthy, J. J. and Smit, D. P. (1984) *South African City: Theory in Analysis and Planning*, Cape Town: Juta.

McCaul, C. (1987) *Satellite in Revolt. KwaNdebele: An Economic and Political Profile*, Johannesburg: South African Institute of Race Relations.

McCormick, Bruce, and Associates (1986) *Pietermaritzburg 2000: Industrial Land Use and Employment*, Consultant Report to the Pietermaritzburg City Engineer.

McGrath, M. D. (1977) *Racial Income Distribution in South Africa*, Durban: University of Natal.

Macoloo, G. C. (1988) 'Housing the urban poor: A case study of Kisumu town, Kenya', *Third World Planning Review*, 10, 2: 159–74.

Mandela, N. (1977) 'Statement during the Rivonia trial, April 20, 1964', in T. G. Karis and G. M. Carter (eds), *From Protest to Challenge: A Documentary History of African Politics in South Africa, 1882–1964*, Stanford University: Hoover Institution Press, pp. 774–91.

Mangin, W. (1967) 'Latin American squatter settlements: a problem and a solution', *Latin American Research Review*, 2: 65–98.

Manona, C. (1985) 'Impact of urbanisation on rural areas: The case of white-owned farms in the Eastern Cape', Paper read at the Annual Conference of South African Anthropologists at the University of Natal, Durban, September.

Maree, J. (1986) 'The past, present and potential role of the democratic trade union movement in South Africa', Paper presented at the conference on 'The Southern African Economy after Apartheid', Centre for Southern African Studies, University of York, 29 September – 2 October.

Markusen, A. (1985) *Profit Cycles, Oligopoly and Regional Development*, Cambridge, Mass.: MIT Press.

Martin, R. (1982) 'The formation of a self-help project in Lusaka', in P. M. Ward (ed.), *Self-Help Housing: A Critique*, London: Mansell Publishing, pp. 251–74.

Massey, D. (1984) *Spatial Divisions of Labour: Social Structures and the Geography of Production*, New York: Methuen.

May, J. (1985) 'Development planning in Transkei – the rural service centre approach', *Working Paper No. 15*, Development Studies Unit, University of Natal, Durban.

Mayo, S. K. (1988) 'Household preferences and expenditure', in L. Rodwin (ed.), *Shelter, Settlement, and Development*, Boston: Allen & Unwin, pp. 61–72.

Mazumdar, D. (1983) 'On the economics of the relative efficiency of small farmers', *Economic Weekly* (special edition), 1 July.

Mera, K. (1973) 'On urban agglomeration and economic efficiency', *Economic Development and Cultural Change*, 21: 309–24.

Meyer, J., Schmenner, R. and Meyer, L. (1980) *Business Location Decisions, Capital Market Imperfections, and the Development of Central City Employment*, Cambridge, Mass.: MIT – Harvard Joint Center for Urban Studies (draft).

Mfuwe Game Lodge (1985) Notes of a meeting chaired by President K. Kaunda between the African National Congress and leading South African capitalists, 13 September.

Misra, B. (1986) 'Public intervention and urban land management', *Habitat International*, 10, 1/2: 65–71.

Moll, T. C. (1986a) 'Review of D. Innes – Anglo American and the rise of modern South Africa', *Transformation*, 1: 138–43.

Moll, T. C. (1986b) '"The art of the possible": macroeconomic policy and income redistribution in Latin America and South Africa', Paper presented at the conference on 'The Southern African Economy after Apartheid', Centre for Southern African Studies, University of York, 29 September – 2 October.

Moller, V. (1988) 'Some thoughts on black urbanization after the abolition of influx control measures', in C. R. Cross and R. J. Haines (eds), *Towards Freehold? Options for Land and Development in South Africa's Black Rural Areas*, Cape Town: Juta, pp. 149–57.

Morobe, M. (1989) 'A perspective on negotiations', Paper delivered to the Transvaal Indian Congress Consultative Conference, 30 July.

Moser, C. O. (1987) 'Introduction', in C. O. Moser and L Peake (eds), *Women, Human Settlements and Housing*, London: Tavistock Publications, pp. 1–11.

Moser, C. and Levy, C. (1986) 'A theory and methodology of gender planning: meeting women's practical and strategic needs', University of London, Development Planning Unit, *Gender and Planning Working Paper No. 11.*

Muller, N. (1984) 'The labour market and poverty in Transkei: Special reference to the changing spatial division of labour', Paper delivered at the Second Carnegie Inquiry into Poverty and Development in Southern Africa, University of Cape Town, 13–19 April.

Muller, N. (1987a) 'Industrial decentralisation: employment creation and urbanisation in Transkei', *Working Paper No. 8*, Development Studies Unit, University of Natal.

Muller, N. (1987b) 'Urbanisation, migration and spatial development in Transkei', Background paper for the National Urbanisation Strategy prepared for Rosmarin, Kriek and Partners, Umtata.

Murray, P. and Szelenyi, I. (1984) 'The city in transition to socialism', *International Journal of Urban and Regional Research*, 8, 1: 90–108.

Musil, J. and Rysavy, Z. (1983) 'Urban and regional processes under capitalism and socialism: a case study from Czechoslavakia', *International Journal of Urban and Regional Research*, 7, 4: 495–527.

National Institute of Transport and Road Research (1985) 'Working document (final draft)', 15 January.

Nattrass, J. (1984) 'Approaches to employment creation in South Africa', *Main Planning Series Report, Volume 54*, Pietermaritzburg: Natal Town and Regional Planning Commission.

Nattrass, J. (1987) 'Politics and liberal economics in the South African context', *South Africa International*, 17, 4: 185–91.

Nattrass, N. (1984) 'Street trading in the Transkei: A struggle against poverty, persecution and prosecution', *Working Paper No. 7*, Development Studies Unit, University of Natal.

Naude, A. H. (1986) 'Urbanisation and transport in the PWV-region: implications of alternative urban structures', Report prepared for the Urban Foundation.

Naude, A. H., Cameron, J. W. M. and Clark, P. M. E. (1987) *Present and Future Settlement, Movement and Migration Patterns in Response to Urbanisation Policies*, Pretoria: Department of Transport.

Nelson, J. (1979) *Access to Power: Politics and the Urban Poor in Developing Nations*, Princeton University Press.

Nidrie, D. (1988a) 'It's testing time for Cosatu', *Work in Progress*, 54: 8–12.

Nidrie, D. (1988b) 'Building on the Freedom Charter', *Work in Progress*, 53: 3–6.

Nolutshungu, S. C. (1982) *Changing South Africa: Political Considerations*, New York: Africana Publishing Company.

O'Brien, C. C. (1986) 'What can become of South Africa?' *The Atlantic Monthly*, March.

Offe, C. (1978) 'The capitalist state and the problem of policy formation', in L. N. Lindbergh *et al*. (eds), *Stress and Contradiction in Modern Capitalism*, Lexington, MA: D. C. Heath, pp. 125–60.

Orkin, M. (1986) *Disinvestment, the Struggle and the Future: What Black South Africans Really Think*, Johannesburg: Ravan Press.

Osmond Lange *et al*. (1982) 'Transkei north-east region – statistical base for planning service centres', Report to the Transkei Government.

Osmond Lange *et al*. (1983a) 'North-east region of Transkei: spatial development plan (1983–2003)', Report to the Transkei Government.

Osmond Lange *et al*. (1983b) 'Statistical base for planning service centres in Transkei's south-east region', Report to the Transkei Government.

Padayachee, A. (1986) 'The politics of international economic relations: South Africa and the International Monetary Fund 1975 and beyond', Paper presented at the conference on 'The Southern African Economy after Apartheid', Centre for Southern African Studies, University of York, 29 September – 2 October.

Page Jr, J. M. and Steel, W. F. (1984) 'Small enterprise development: economic issues from African experience', *World Bank Technical Paper No. 29*, Washington, DC.

Payne, G. K. (1982) 'Self-help housing: a critique of the gecekondus of Ankara', in P. M. Ward (ed.), *Self-Help Housing: A Critique*, London: Mansell Publishing, pp. 117–39.

Payne, G. K. (1984) 'Introduction', in G. K. Payne (ed.), *Low-Income Housing in the Developing World: The Role of Site and Services and Settlement Upgrading*, Chichester: John Wiley, pp. 1–16.

Peattie, L. (1987) 'Shelter, development, and the poor', in L. Rodwin (ed.), *Shelter, Settlement, and Development*, Boston: Allen & Unwin, pp. 263–80.

Perlman, J. E. (1976) *The Myth of Marginality: Urban Poverty and Politics in Rio de Janeiro*, University of California Press.

Phillips, M. (1989) 'It takes two to tango', *Work in Progress*, 60: 10–18.

Pillay, P. N. (1984) 'Poverty in the Pretoria–Witwatersrand–Vereeniging area: a survey of research', Paper presented at the Second Carnegie Inquiry into Poverty and Development in Southern Africa, University of Cape Town, 13–19 April.

Please, S. and Amoako, K. Y. (1986) 'OAU, ECA and the World Bank: Do they really disagree?' in J. Ravenhill (ed.), *Africa in Economic Crisis*, New York: Columbia University Press, pp. 127–48.

Prior, A. (1983/4) 'South African exile politics: a case study of the African National Congress and the South African Communist Party', *Journal of Contemporary African Studies*, 3, 1/2: 181–96.

Progressive Federal Party (1986) 'There is a way to save SA' (pamphlet).

Ram, R. (1988) 'Economic development and income inequality: Further evidence on the U-curve hypothesis', *World Development*, 16, 11: 1371–6.

Ramaphosa, C. (1987) 'Cyril Ramaphosa on the NUM Congress', *South African Labour Bulletin*, 12, 3: 45–56.

Rao, D. C. (1974) 'Urban target groups', in H. Chenery *et al.* (ed.), *Redistribution with Growth*, Oxford University Press, published for the World Bank, pp. 136–58.

Renaud, B. (1979) *National Urbanization Policies in Developing Countries*, Staff Working Paper No. 347, Washington DC: World Bank.

Renaud, B. (1987) 'Financing shelter', in L. Rodwin (ed.), *Shelter, Settlement, and Development*, Boston: Allen & Unwin, pp. 179–203.

Richardson, G. B. (1971) 'Planning versus competition', *Soviet Studies*, 12, 3: 433–47.

Richardson, H. W. (1978) 'Growth centres, rural development and national urban policy', *International Regional Science Review*, 3, 2: 133–52.

Richardson, H. W. (1987) 'Spatial strategies, the settlement pattern, and shelter and service policies', in L. Rodwin (ed.), *Shelter, Settlement, and Development*, Boston: Allen & Unwin, pp. 207–35.

Riordan, R. (1988) 'The Freedom Charter and the nationalisation of industry', Paper presented at the Freedom Charter Regional Consultative Conference, convened by Idasa, Port Elizabeth, 14 May.

Robinson, P. S. (1981) 'Development planning in Transkei – a summary of the Hawkins Report', *Development Studies Southern Africa*, 3, 4: 381–418.

Roddick, J. E. (1976) 'Class structure and class politics in Chile', in P. J. O'Brien (ed.), *Allende's Chile*, Praeger, pp. 1–26.

Rodwin, L. and Sanyal, B. (1987) 'Shelter, settlement and development', in L. Rodwin (ed.), *Shelter, Settlement, and Development*, Boston: Allen & Unwin, pp. 3–31.

Rondinelli, D. A. (1983) *Secondary Cities in Developing Countries: Policies for Diffusing Urbanisation*, Beverly Hills, CA: Sage.

Ronen, D. (1986) 'Ethnicity, politics and development: an introduction', in D. L. Thompson and D. Ronen (eds), *Ethnicity, Politics and Development*, Boulder, Colorado: Lynne Rienner Publishers, pp. 1–10.

Roukens de Lange, A. (1984) 'Demographic tendencies, technological development and the future of apartheid', Paper delivered at the conference on 'Economic Development and Racial Domination', University of the Western Cape, 8–10 October.

Sampson, A. (1987) *Black and Gold: Tycoons, Revolutionaries and Apartheid*, London: Hodder & Stoughton.

Sarakinsky, I. (1988) 'State, strategy and transition in South Africa: Historical and contemporary perspectives', Paper presented at a seminar of the African Studies Institute, University of the Witwatersrand, 8 August.

Sattherwaite, D. (1983) 'Public land acquisition and shelter of the poor', in 'Report of the United Nations Seminar of Experts on Land for Housing the Poor', Tallberg and Stockholm, March. Stockholm: Swedish Council for Building Research, pp. 49–60.

Saul, J. S. (1986) 'Introduction: the revolutionary prospect', in J. S. Saul and A. Gelb, *The Crisis in South Africa*, New York: Monthly Review Press.

Schlemmer, L. (1985) 'Squatter communities: safety valves in the rural–urban nexus', in H. Giliomee and L. Schlemmer, *Up against the Fences: Poverty, Passes and Privilege in South Africa*, Cape Town: David Philip, pp. 167–91.

Senior, B. (1984) 'Factors affecting residential density: a search for the Zen of density', Doctoral thesis in the Faculty of Architecture, University of the Witwatersrand, Johannesburg.

Shaw, T. M. (1986) 'The African crisis: debates and dialectics over alternative development strategies for the continent', in J. Ravenhill (ed.), *Africa in Economic Crisis*, New York: Columbia University Press, pp. 108–26.

Shoup, D. C. (1983) 'The rationale for government intervention', in H. B. Dunkerley (ed.), *Urban Land Policy: Issues and Opportunities*, Oxford University Press, published for the World Bank, pp. 132–52.

Simkins, C. E. W. (1979) 'The distribution of personal income among recipients in South Africa, 1970 and 1976', Durban: University of Natal.

Simkins, C. E. W. (1981) 'Agricultural production in the African reserves of South Africa, 1918–1969', *Journal of Southern African Studies*, 7, 2: 256–83.

Simkins, C. E. W. (1982) 'Structural unemployment revisited', University of Cape Town, *Saldru Fact Sheet No. 1*.

Simkins, C. E. W. (1983) *Four Essays on the Past, Present and Possible Future of the Distribution of the Black Population of South Africa*, University of Cape Town: Southern Africa Labour and Development Research Unit.

Simkins, C. E. W. (1984) 'Public expenditure and the poor: Political and economic constraints on policy choices up to the year 2000', Paper delivered at the Second Carnegie Inquiry into Poverty and Development in Southern Africa (Paper No. 253), University of Cape Town, 13–19 April.

Simkins, C. E. W. (1985) 'Projecting the spatial distribution of black employment and population in South Africa in the year 2000', Confidential report prepared for the Urban Foundation.

Simkins, C. E. W. (1986) 'How much socialism will be needed to end poverty in South Africa?' Paper presented at the conference on 'The Southern African Economy after Apartheid', Centre for Southern African Studies, University of York, 29 September – 2 October.

Simkins, C. E. W. (1990) 'Population Policy' in R. Schrire (ed.), *Critical Choices for South Africa: An Agenda for the 1990s*, Cape Town: Oxford University Press, pp. 215–33.

Sklar, R. T. (1975) *Corporate Power in an African State: The Political Impact of Multinational Mining Companies in Zambia*, Berkeley: University of California Press.

Slabbert, F. van Zyl (1986) 'Haphazard holding actions', *Star*, 25 April.

Slabbert, F. van Zyl and Welsh, D. (1979) *South Africa's Options: Strategies for Sharing Power*, New York: St Martin's Press.

Smith, J. D. and Coetzee, S. F. (1987) *KwaZulu's Potential to Attract Specific Types of Industries*, Johannesburg: Development Bank of Southern Africa.

Smith, R. S. (1979) 'The effects of land taxes on development timing and rates of change in land prices', in R. W. Bahl (ed.), *The Taxation of Urban Property in Less Developed Countries*, University of Wisconsin Press, pp. 137–62.

South Africa (1987) *Present and Future Settlement, Movement and Migration Patterns in Response to Urbanisation Policies*, VV 1/87, Pretoria: Department of Transport.

South Africa, Republic of (1916) *Report of the Native Lands Commission* (Beaumont Commission), UG 19, Pretoria: Government Printer.

South Africa, Republic of (1955) *Summary of the Report of the Commission for the Socio-Economic Development of the Bantu Areas within the Union of South Africa* (Tomlinson Commission), UG 61, Pretoria: Government Printer.

South Africa, Republic of (1970) *The Second Report of the Commission of Inquiry into Agriculture*, RP34, Pretoria: Government Printer.

South Africa, Republic of (1981a), Office of the Prime Minister – Physical Planning Branch, *A Spatial Development Strategy for the PWV Complex*, Pretoria: Government Printer.

South Africa, Republic of (1981b) *Census of Agriculture 1981*, Pretoria: Central Statistical Services Report No. 06–0117.

South Africa, Republic of (1984a) *First Report of the Commission of Inquiry into Township Establishment and Related Matters* (Venter Commission), RP20, Pretoria: Government Printer.

South Africa, Republic of (1984b) *Report of the Committee for Economic Affairs of the President's Council on Measures which Restrict the Functioning of a Free Market Orientated System in South Africa* (Raubenheimer Commission), P.C.1, Cape Town: Government Printer.

South Africa, Republic of (1985) *Report of the Committee for Constitutional Affairs of the President's Council on an Urbanisation Strategy for the Republic of South Africa*, P.C.3, Cape Town: Government Printer.

South Africa, Republic of (1986a), Department of Constitutional Development and Planning, *Draft Guide Plan for the Central Witwatersrand*, Pretoria: Government Printer.

South Africa, Republic of (1986b) *White Paper on Urbanisation*, W.P., Pretoria: Government Printer.

South African Institute of Race Relations (1983) *Race Relations Survey 1982*, Johannesburg.

South African Institute of Race Relations (1985) *Race Relations Survey 1984*, Johannesburg.

South African Institute of Race Relations (1987) *Race Relations Survey 1987*, Johannesburg.

South African Labour Bulletin (1987) 'A message to all members of COSATU', 12, 2: 48–54.

Southall, R. (1986) 'South Africa: constraints on socialism', Paper presented at the conference on 'The Southern African Economy after

Apartheid', Centre for Southern African Studies, University of York, 29 September – 2 October.

Southey, C. (1982) 'Land tenure in the Transkei', University of the Transkei, *IMDS Discussion Paper No. 9*.

Spies, P. H. (n.d.) 'The impact of dominant societal forces on South African agriculture', Institute for Futures Research, University of Stellenbosch.

Stadler, J. J., Du Pisanie, J. A. and Kritzinger, L. (1984) *The Logic of the Federal Option*, Johannesburg: Mercabank Report.

Stanback, T., Bearse, P. J., Noyelle, T. J., Karasek, R. A. (1983) *Services: The New Economy*, Totowa, NJ: Rowman & Allanheld.

Stanwix, J. (1985) 'A study of the Natal regional economy', *Natal Town and Regional Planning Report Vol. 66*, Pietermaritzburg: Natal Town and Regional Planning Commission.

Stavrou, S. E. (1987) 'The restructuring of agrarian capitalism after 1950', Paper presented at the workshop on 'The South African Agrarian Question: Past, Present and Future', University of the Witwatersrand, 22–24 May.

Stohr, W. and Todtling, F. (1978) 'An evaluation of regional policies – experiences in market and mixed economies', in N. M. Hansen (ed.), *Human Settlement Systems*, Cambridge, Mass.: Ballinger, pp. 85–119.

Sunter, C. (1987) *The World and South Africa in the 1990s*, Tafelberg: Human and Rousseau.

Surplus People's Project (n.d. [1983]) *Forced Removals in South Africa. Vol. 2. The Eastern Cape*, published by the Project.

Szelenyi, I. (1983) *Urban Inequalities under State Socialism*, New York: Oxford University Press.

Tapscott, C. P. C., Haines, R. J. and Wakelin, P. M. (1984) 'A critique of rural service centres, with special reference to Transkei', IMDS, University of the Transkei, Umtata.

Thomas, W. (1983) 'Socio-economic development in Transkei', Extra-Mural Studies, University of Cape Town.

Thomas, W. (1984) 'Financing rural development – with particular reference to Transkei', Paper delivered at the Second Carnegie Inquiry into Poverty and Development in Southern Africa, University of Cape Town, 13–19 April.

Threlfall, M. (1976) 'Shantytown dwellers and people's power', in P. J. O'Brien (ed.), *Allende's Chile*, Praeger, pp. 167–91.

Tipple, A. G. (1976) 'The low cost housing market in Kitwe, Zambia', *Ekistics*, 41 (March): 148–52.

Tomlinson, R. (1982) 'The political economy of regional inequality in post-independence Kenya', *South African Geographical Journal*, 64, 2: 21–40.

Tomlinson, R. (1983a) 'Industrial decentralisation and the relief of poverty in the homelands', *South African Journal of Economics*, 51, 4: 544–63.

Tomlinson, R. (1983b) 'On "Rethinking regional inequality: the case of Kenya"', *South African Geographical Journal*, 65, 2: 197–200.

Tomlinson, R. (1984) 'A development framework for Gazankulu', *Development South Africa*, 1, 1: 116–22.

Tomlinson, R. (1986) 'From decentralisation concessions to a tax-free zone: the case of Ciskei', MBA thesis in the Graduate School of Business Administration, University of the Witwatersrand, Johannesburg.

Tomlinson, R. (1988a) 'South Africa's urban policy: a new form of influx control', *Urban Affairs Quarterly*, 23, 4: 487–510.

Tomlinson, R. (1988b) 'South Africa: competing images of the post-apartheid state', *African Studies Review*, 31, 2: 35–60.

Tomlinson, R. and Addleson, M. (1985) 'Ciskei as a free enterprise zone', *Development Southern Africa*, 2, 2: 174–86.

Tomlinson, R. and Addleson, M. (1986) 'Trends in industrial decentralisation: a review of Bell's hypothesis', *South African Journal of Economics*, 54, 4: 381–94.

Tomlinson, R. and Addleson, M. (1987) 'Is the state's regional policy in the interests of capital?' in R. Tomlinson and M. Addleson (eds), *Regional Restructuring under Apartheid: Contemporary Perspectives on Urban and Regional Policies in South Africa*, Johannesburg: Ravan Press, pp. 55–73.

Trachte, K. and Ross, R. (1985) 'The crisis of Detroit and the emergence of global capitalism', *International Journal of Urban and Regional Research*, 9, 2: 187–216.

Transkei, 'Republic of' (1980) *Development Strategy 1980–2000*, Umtata.

Transkei, 'Republic of' (1983) *Development Priorities and Public Sector Spending 1983–1988* (White Paper), Umtata.

Transkei, 'Republic of' (1987) 'A strategy for promoting dryland crop production through the establishment of comprehensive farmer support programmes', Department of Agriculture and Forestry and TRACOR, Umtata.

Transvaal Rural Action Committee (1985) *Newsletter No. 9* (September).

Turner, J. F. C. (1968) 'Housing priorities, settlement patterns and urban development in modernizing countries', *Journal of the American Institute of Planners*, 34: 355–60.

Turner, J. F. C. (1969) *Freedom to Build*, London: Macmillan.

Uhlig, M. (1986) 'Inside the African National Congress', *New York Times Magazine*, 12 October.

United Nations (1976) 'Habitat: Conference on Settlements', Vancouver, 31 May – 11 June.

United Nations (1983) 'Report of the United Nations Seminar of Experts on Land for Housing the Poor', Tallberg and Stockholm, March. Stockholm: Swedish Council for Building Research, pp. 5–47.

United Nations Centre for Human Settlements (HABITAT) (1981) *Report on the ad hoc Expert Group Meeting on the Development of the Indigenous Construction Center*, 23–30 November, Nairobi.

Van der Berg, S. (1986) 'Forecasts of inter-racial income distribution to the year 2000', *Occasional Paper No. 9*, Institute for Futures Research, University of Stellenbosch, March.

Van Heerden, D. (1986) 'The new Nats', *Frontline*, 6, 2: 35–7.

Van Huyck, A. P. (1982) 'Land tenure choices in urban upgrading projects within the context of national land policy', edited by A. P. Van Huyck from the writings of Professor W. Doebele, Harvard University (mimeo).

Van Huyck, A. P. (1987) 'Defining the roles of the public and private sectors in urban development', Paper prepared for the 'Regional Seminar on Major National Urban Policy Issues', Sponsored by the Asian Development Bank and the United Nations Centre for Regional Development, 3–7 February.

Vaughan, R. and Bearse P. (1981) 'Federal economic development programs: a framework for design and evaluation', in R. Friedman and W. Schweker (eds), *Expanding the Opportunities to Produce*, Washington, DC: Corporation for Enterprise Development.

Wakelin, P. M. (1983) 'Migrant labour in Transkei', *IMDS Statistical Series 2–83, University of Transkei, Umtata*.

Ward, P. M. (1982) 'Introduction and purpose', in P. M. Ward (ed.), *Self-Help Housing: A Critique*, London: Mansell Publishing, pp. 1–14.

Webster, E. (1986) 'The goals of management and labour – industrial relations in a post-apartheid economy', Paper presented at the conference on 'The Southern African Economy after Apartheid', Centre for Southern African Studies, University of York, 29 September – 2 October.

Weiner, D., Moyo, S., Munslow, B. and O'Keefe, P. (1985) 'Land use and agricultural productivity in Zimbabwe', *Journal of Modern African Studies*, 23, 2: 251–85.

Wellings, P. (1985) 'Review of "Regions in Question: Space, Development Theory and Regional Policy"', Charles Gare in *Development Southern Africa*, 2, 3: 422–27, London: Methuen.

Wellings, P. and Black, A. (1986) 'Industrial decentralisation under apartheid: the relocation of industry to the South Africa periphery', *World Development*, 14, 1: 1–38.

Wellings, P. and Black, A. (1987) 'Industrial decentralisation under apartheid: an empirical assessment', in R. Tomlinson and M. Addleson (eds), *Regional Restructuring under Apartheid: Contemporary Perspectives on Urban and Regional Policies in South Africa*, Johannesburg: Ravan Press, pp. 181–206.

Western, J. (1982) 'The geography of urban social control: Group Areas and the 1976 and 1980 civil unrest in Cape Town', in D. M. Smith (ed.), *Living under Apartheid*, London: Allen & Unwin, pp. 215–29.

Williamson, J. G. (1965) 'Regional inequality and the process of national development: a description of the patterns', *Economic Development and Cultural Change*, 13, 4, pt 2: 3–45.

Wilson, F. and Ramphele, M. (1989) *Uprooting Poverty: The South African Challenge*, Cape Town: David Philip.

Wolpe, H. (1972) 'Capitalism and cheap labour power in South Africa: from segregation to apartheid', *Economy and Society*, 1: 424–56.

Yawitch, J. (1981) *Betterment: The Myth of Homeland and Agriculture*, Johannesburg: South African Institute of Race Relations.

Yudelman, D. (1987) 'State and capital in contemporary South Africa', in J. Butler, R. Elphick and D. Welsh (eds), *Democratic Liberalism in South Africa: Its History and Prospect*, Middletown, Conn.: Wesleyan University Press, pp. 250–68.

Zille, H. (1983) 'Restructuring the industrial decentralisation strategy', in South African Research Services (ed.), *South African Review: Same Foundation, New Facades?*, Johannesburg: Ravan Press, pp. 58–71.

Index

Printed in the United States
by Baker & Taylor Publisher Services